Chapter 12

Two-Dimensional Shapes

Made in the United States
Text printed on 100%
recycled paper

Printed in the U.S.A.

ISBN 978-0-544-34218-7

2 3 4 5 6 7 8 9 10 0928 22 21 20 19 18 17 16 15 14

4500476012 ^ B C D E F G

Dear Students and Families,

Welcome to **Go Math!**, Grade 3! In this exciting mathematics program, there are hands-on activities to do and real-world problems to solve. Best of all, you will write your ideas and answers right in your book. In **Go Math!**, writing and drawing on the pages helps you think deeply about what you are learning, and you will really understand math!

By the way, all of the pages in your **Go Math!** book are made using recycled paper. We wanted you to know that you can Go Green with **Go Math!**

Sincerely,

The Authors

Made in the United States
Text printed on 100% recycled paper

Authors

Juli K. Dixon, Ph.D.
Professor, Mathematics Education
University of Central Florida
Orlando, Florida

Edward B. Burger, Ph.D.
President, Southwestern University
Georgetown, Texas

Steven J. Leinwand
Principal Research Analyst
American Institutes for Research (AIR)
Washington, D.C.

Contributor

Rena Petrello
Professor, Mathematics
Moorpark College
Moorpark, California

Matthew R. Larson, Ph.D.
K-12 Curriculum Specialist for Mathematics
Lincoln Public Schools
Lincoln, Nebraska

Martha E. Sandoval-Martinez
Math Instructor
El Camino College
Torrance, California

English Language Learners Consultant

Elizabeth Jiménez
CEO, GEMAS Consulting
Professional Expert on English Learner Education
Bilingual Education and Dual Language
Pomona, California

Geometry

Critical Area

Critical Area Describing and analyzing two-dimensional shapes

GO DIGITAL

Go online! Your math lessons are interactive. Use *i*Tools, Animated Math Models, the Multimedia eGlossary, and more.

Chapter 12 Overview

In this chapter, you will explore and discover answers to the following **Essential Questions**:

- What are some ways to describe and classify two-dimensional shapes?
- How can you describe the angles and sides in polygons?
- How can you use sides and angles to describe quadrilaterals and triangles?
- How can you use properties of shapes to classify them?
- How can you divide shapes into equal parts and use unit fractions to describe the parts?

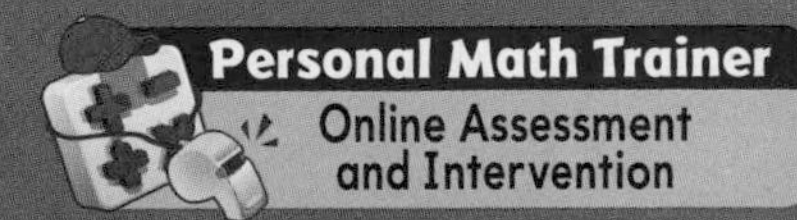

CRITICAL AREA REVIEW PROJECT MAKE A MOSAIC: *www.thinkcentral.com*

Practice and Homework

Lesson Check and Spiral Review in every lesson

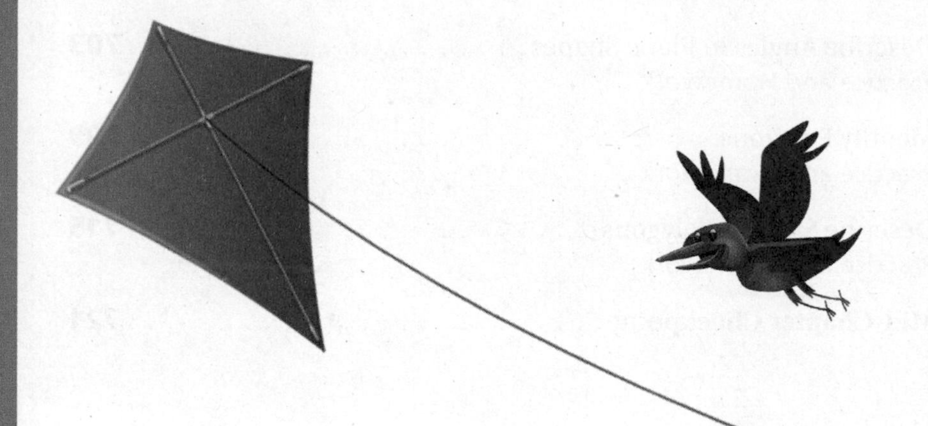

Critical Area

Geometry

Common Core

CRITICAL AREA Describing and analyzing two-dimensional shapes

Students at Dommerich Elementary helped design and construct a mosaic to show parts of their community and local plants and animals.

Make a Mosaic

Have you ever worked to put puzzle pieces together to make a picture or design? Pieces of paper can be put together to make a colorful work of art called a mosaic.

Get Started

Materials ■ construction paper ■ glue ■ ruler ■ scissors

Work with a partner to make a paper mosaic. Use the Important Facts to help you.

- Draw a simple pattern on a piece of paper.
- Cut out shapes, such as rectangles, squares, and triangles of the colors you need from construction paper. The shapes should be about 1 inch on each side.
- Glue the shapes into the pattern. Leave a little space between each shape to make the mosaic effect.

Describe and compare the shapes you used to make your mosaic.

Important Facts

- Mosaics is the art of using small pieces of materials, such as tiles or glass, to make a colorful picture or design.
- Mosaic pieces can be small plane shapes, such as rectangles, squares, and triangles.
- Mosaic designs and patterns can be anything from simple flower shapes to common objects found in your home or patterns in nature.

____

____

____

____

____

Completed by _______________________________

Two-Dimensional Shapes

Show What You Know

Check your understanding of important skills.

Name ______________________________

▶ Plane Shapes

1. Color the triangles blue.

 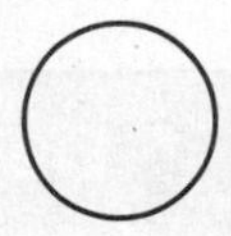 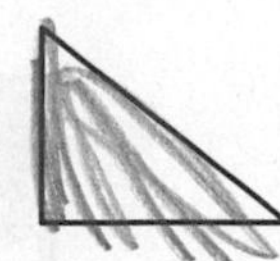 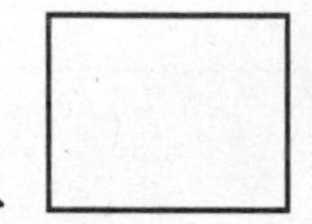

2. Color the rectangles red.

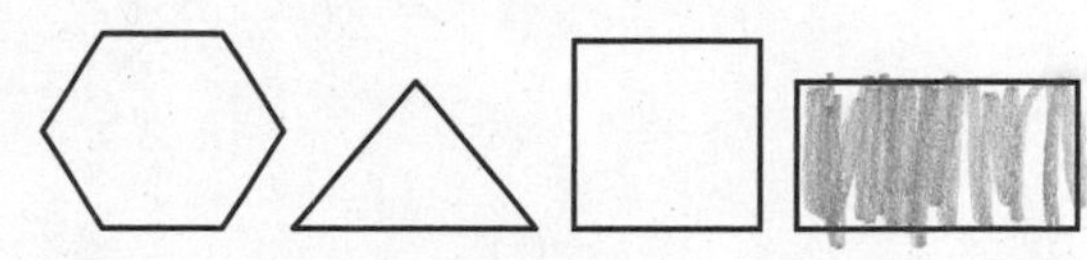

▶ Number of Sides **Write the number of sides.**

3. 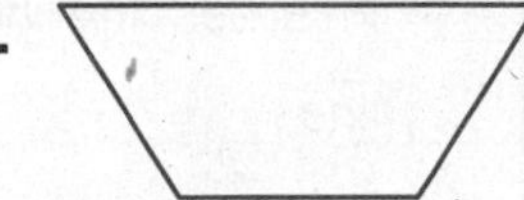

__4__ sides

4. 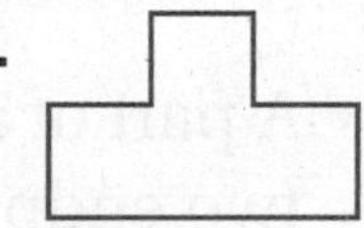

__6__ sides

5. Circle the shapes that have 4 or more sides.

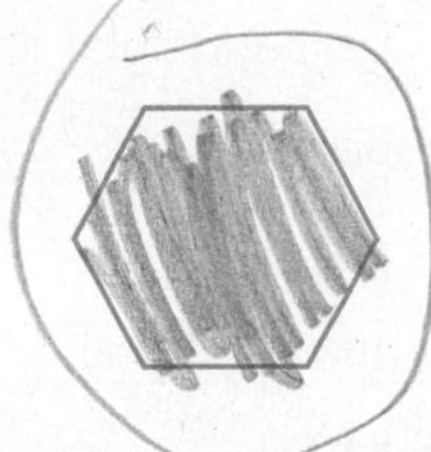  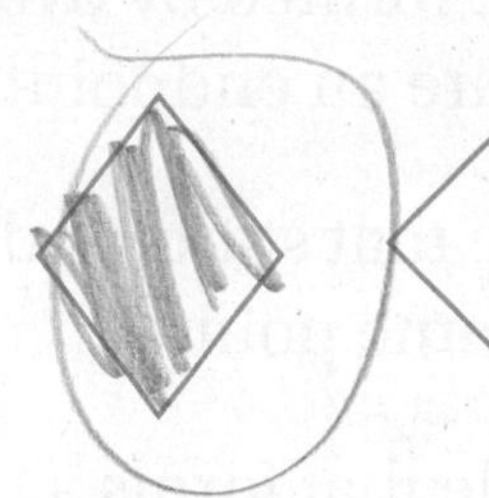 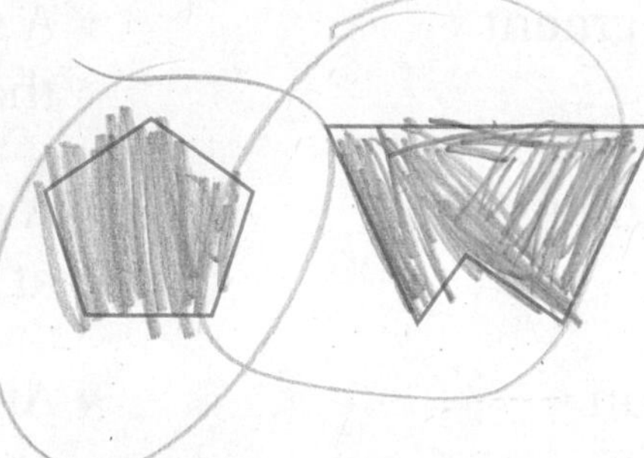

Whitney found this drawing that shows 9 small squares. Have students find larger squares in the drawing. How many squares are there in all? Explain.

Vocabulary Builder

▶ Visualize It

Complete the tree map by using the words with a ✓.

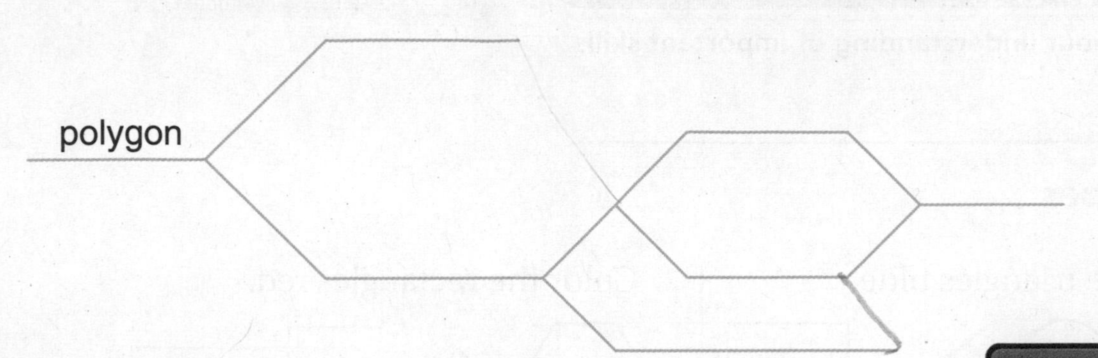

▶ Understand Vocabulary

Draw a line to match the word with its definition.

1. closed shape •	• A part of a line that includes two endpoints and all the points between them
2. line segment •	• A shape formed by two rays that share an endpoint
3. right angle •	• A shape that starts and ends at the same point
4. hexagon •	• An angle that forms a square corner
5. angle •	• A closed plane shape made up of line segments
6. polygon •	• A polygon with 6 sides and 6 angles

Preview Words

- angle
- closed shape
- hexagon
- intersecting lines
- line
- line segment
- open shape
- parallel lines
- perpendicular lines
- point
- polygon
- ✓ quadrilateral
- ray
- ✓ rectangle
- ✓ rhombus
- right angle
- ✓ square
- ✓ trapezoid
- ✓ triangle
- Venn diagram
- vertex

- Interactive Student Edition
- Multimedia eGlossary

Chapter 12 Vocabulary

English	Spanish	Number
angle	ángulo	2
endpoint	extremo	19
intersecting lines	líneas secantes	36
line	línea	42
line segment	segmento	44
octagon	octágono	54
parallel lines	líneas paralelas	56
pentagon	pentágono	58

The point at either end of a line segment

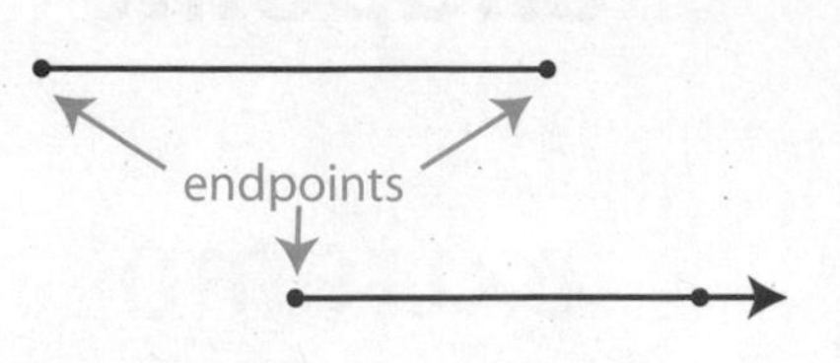

A shape formed by two rays that share an endpoint

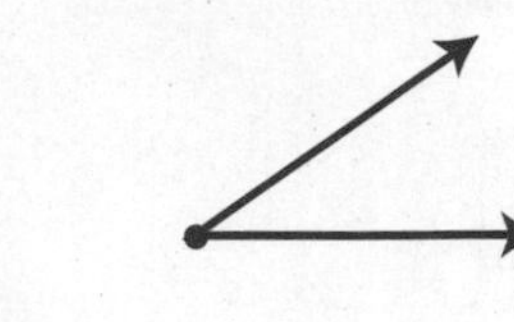

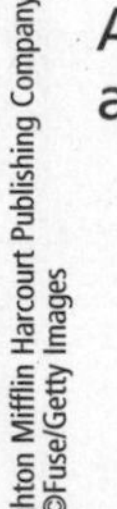

A straight path extending in both directions with no endpoints

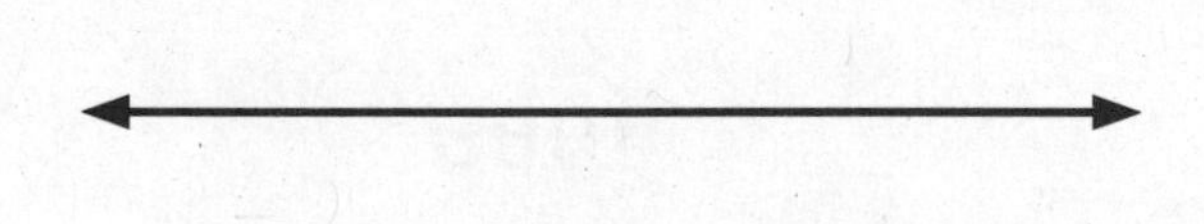

Lines that meet or cross

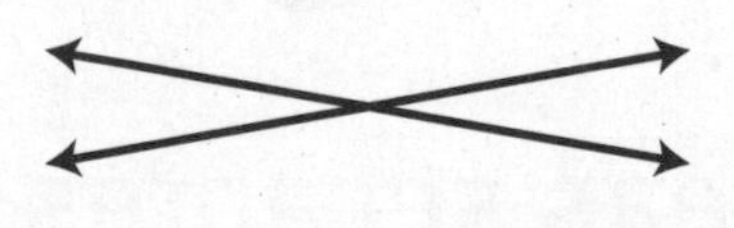

A polygon with eight sides and eight angles

A part of a line that includes two points, called endpoints, and all of the points between them

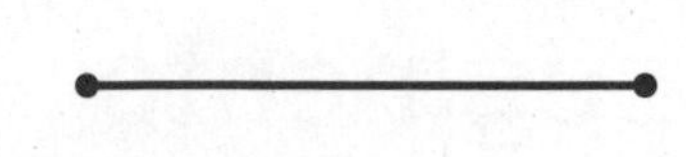

A polygon with five sides and five angles

Lines in the same plane that never cross and are always the same distance apart

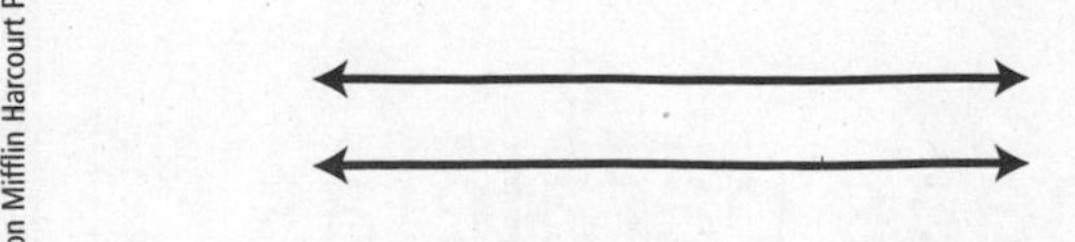

perpendicular lines

líneas perpendiculares

60

polygon

polígono

64

quadrilateral

cuadrilátero

66

ray

semirrecta

68

rhombus

rombo

71

trapezoid

trapecio

78

Venn diagram

diagrama de Venn

81

vertex

vértice

82

A closed plane shape with straight sides that are line segments

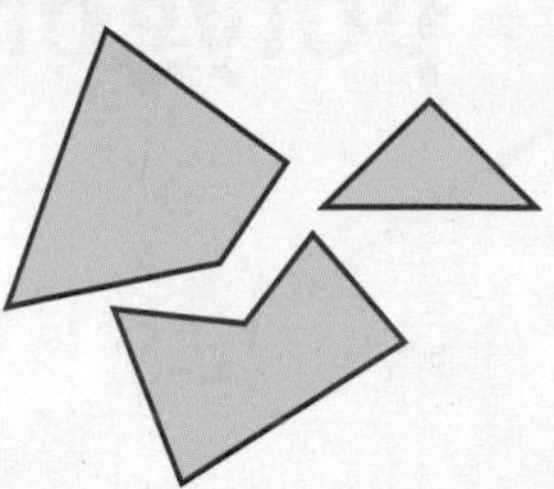

Lines that intersect to form right angles

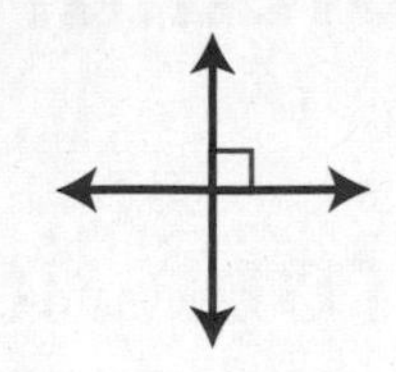

A part of a line, with one endpoint, that is straight and continues in one direction

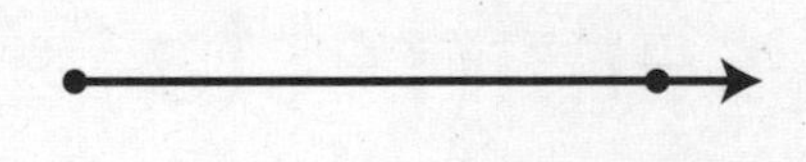

A polygon with four sides and four angles

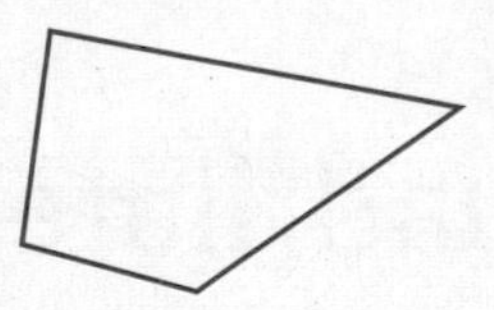

A quadrilateral with exactly one pair of parallel sides

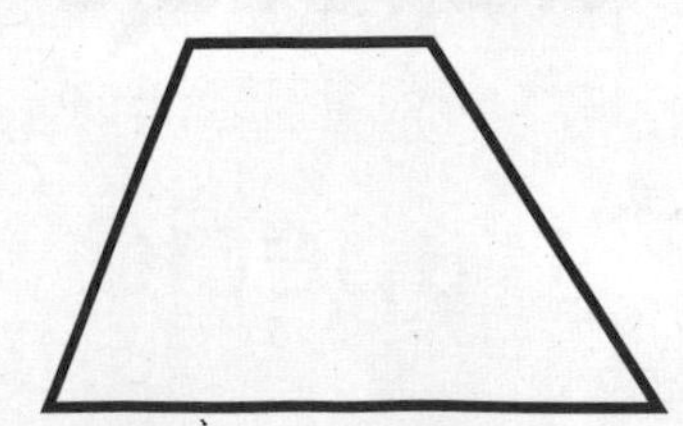

A quadrilateral with two pairs of parallel sides and four sides of equal length

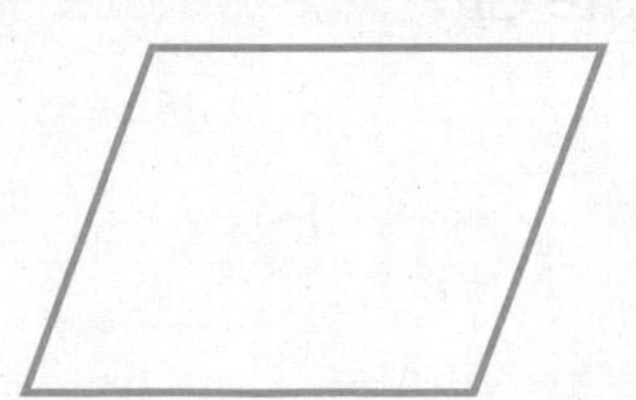

The point at which two rays of an angle or two (or more) line segments meet in a plane shape or where three or more edges meet in a solid shape

Examples:

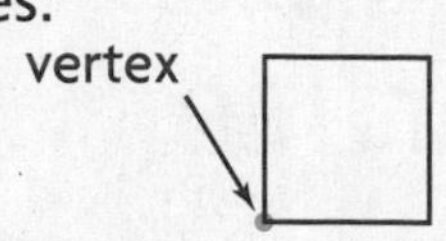

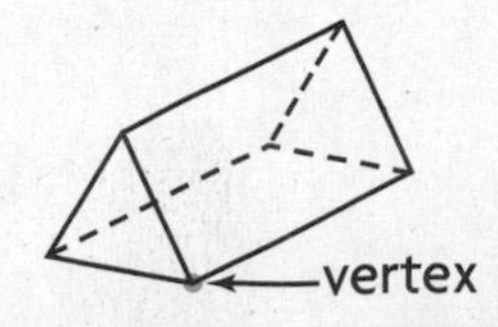

A diagram that shows relationships among sets of things

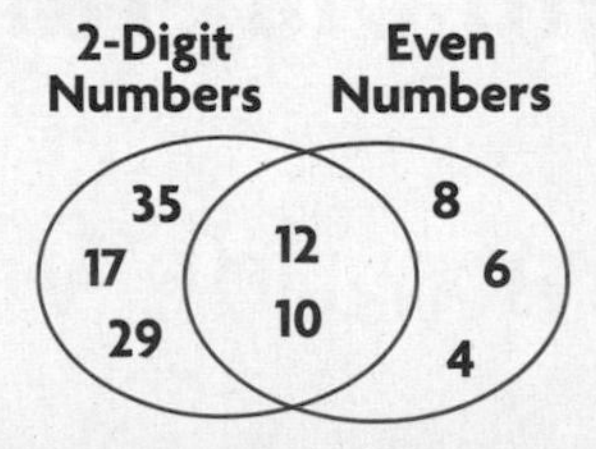

Going to an Art Museum

For 2 players

Materials

- 1 red connecting cube
- 1 blue connecting cube
- 1 number cube
- Clue Cards

How to Play

1. Choose a connecting cube and put it on START.
2. Toss the number cube to take a turn. Move your connecting cube that many spaces.
3. If you land on one of these spaces:
   Blue Space Follow the directions printed in the space.
   Red Space Take a Clue Card from the pile. If you answer the question correctly, keep the Clue Card. If you do not, return the Clue Card to the bottom of the pile.
4. Collect at least 5 Clue Cards. Move around the track as many times as you need to.
5. Only when you have 5 Clue Cards, you follow the closest center path to reach FINISH.
6. The first player to reach FINISH wins.

Word Box
angle
endpoints
intersecting lines
line
line segment
octagon
parallel lines
pentagon
perpendicular lines
polygon
quadrilateral
ray
rhombus
trapezoid
Venn diagram
vertex

TAKE A CLUE CARD

Your favorite artist's work is on display. Move ahead 1.

FINISH

TAKE A CLUE CARD

You stand too close to a painting and set off the alarms. Go back 1.

The museum is closed today. Go back 1.

TAKE A CLUE CARD

Free tours today! Move ahead 1.

FINISH

TAKE A CLUE CARD

START ▶

The Write Way

Reflect

Choose one idea. Write about it.

- In two minutes, draw and label as many examples of polygons that you can. Use a separate piece of paper for your drawing.
- Work with a partner to explain and illustrate parallel, intersecting, and perpendicular lines. Use a separate piece of paper for your drawing.
- A reader of your math advice column writes, “I confuse rhombuses, squares, rectangles, and trapezoids. How can I tell the difference between these quadrilaterals?” Write a letter to your reader offering step-by-step advice.

Name ____________________

Describe Plane Shapes

Essential Question What are some ways to describe two-dimensional shapes?

Common Core Geometry—3.G.A.1

MATHEMATICAL PRACTICES
MP5, MP6, MP7

Unlock the Problem

An architect draws plans for houses, stores, offices, and other buildings. Look at the shapes in the drawing at the right.

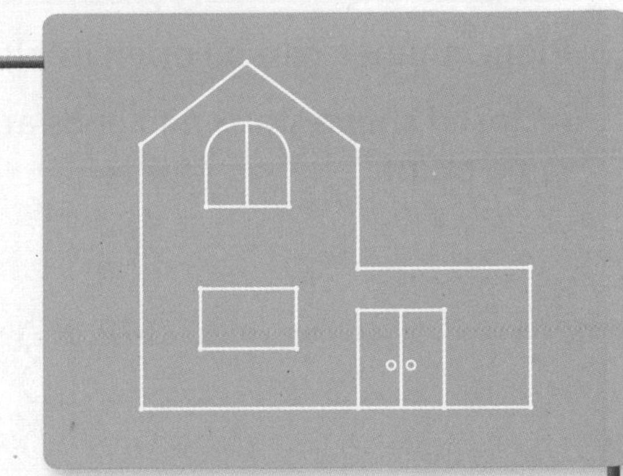

A **plane shape** is a shape on a flat surface. It is formed by points that make curved paths, line segments, or both.

point
- is an exact position or location

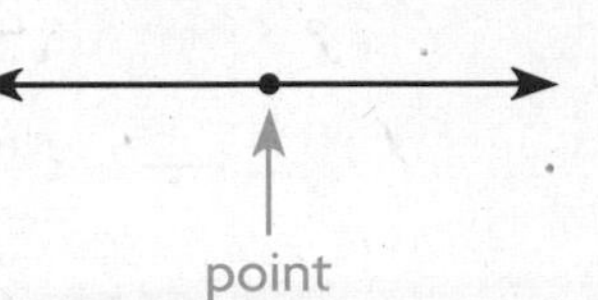

line
- is a straight path
- continues in both directions
- does not end

endpoints
- points that are used to show segments of lines

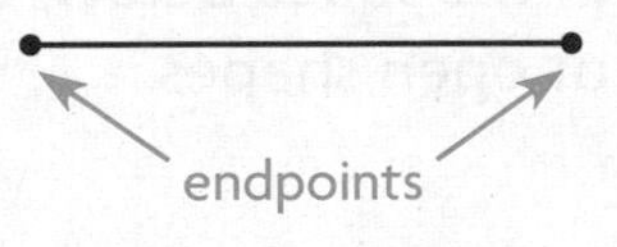

line segment
- is straight
- is part of a line
- has 2 endpoints

ray
- is straight
- is part of a line
- has 1 endpoint
- continues in one direction

Some plane shapes are made by connecting line segments at their endpoints. One example is a square. Describe a square using math words.

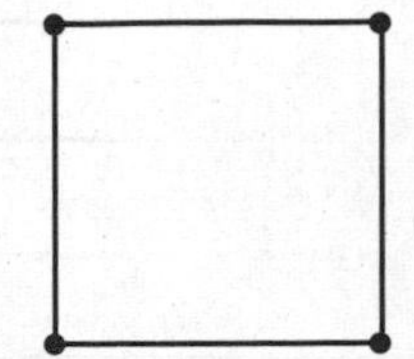

Think: How many line segments and endpoints does a square have?

A square has ____ line segments. The line segments meet only at their ____________.

Math Talk MATHEMATICAL PRACTICES 3

Apply Why can you not measure the length of a line?

Plane shapes have length and width but no thickness, so they are also called **two-dimensional shapes**.

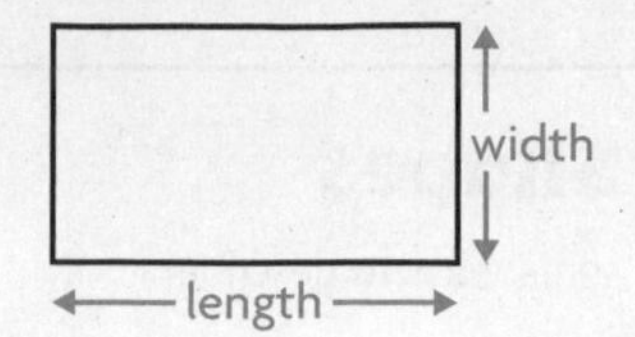

Try This! Draw plane shapes.

Plane shapes can be open or closed.

A **closed shape** starts and ends at the same point.

In the space below, draw more examples of closed shapes.

An **open shape** does not start and end at the same point.

In the space below, draw more examples of open shapes.

Math Talk MATHEMATICAL PRACTICES 6

Explain whether a shape with a curved path must be a closed shape, an open shape, or can be either.

- Is the plane shape at the right a closed shape or an open shape? Explain how you know.

It's closed because it's connected and closed.

Name ______________________

Share and Show

1. Write how many line segments the shape has. 3

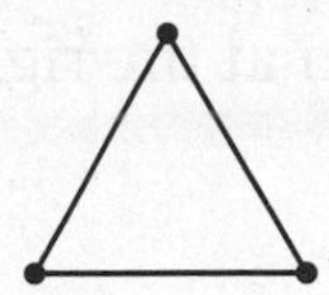

Circle all the words that describe the shape.

2. 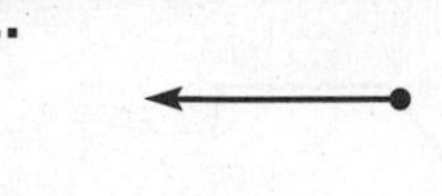

ray

point

3. 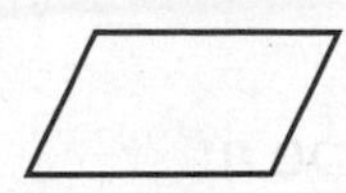

open shape

closed shape

4. 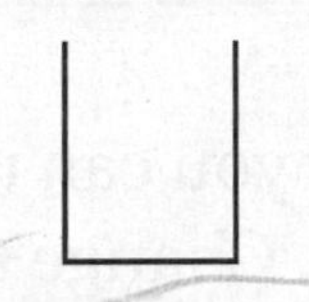

open shape

closed shape

5. 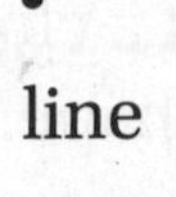

line

line segment

Write whether the shape is *open* or *closed*.

6. 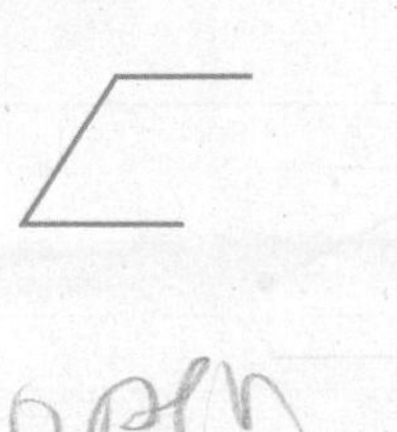

open

7. 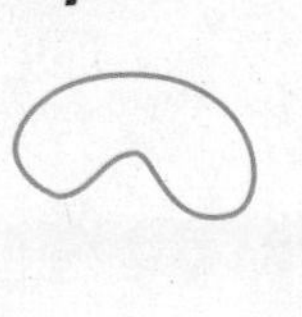

closed

8. 

closed

9. 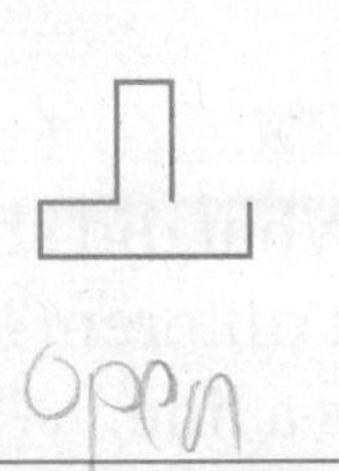

open

Math Talk MATHEMATICAL PRACTICES 1

Describe How do you know whether a shape is open or closed?

On Your Own

Write how many line segments the shape has.

10. 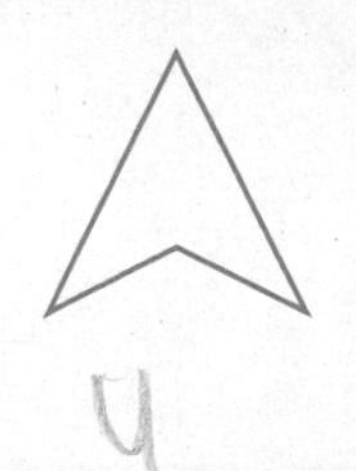

4 line segments

11. 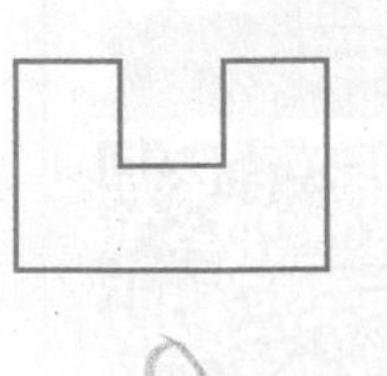

8 line segments

12. 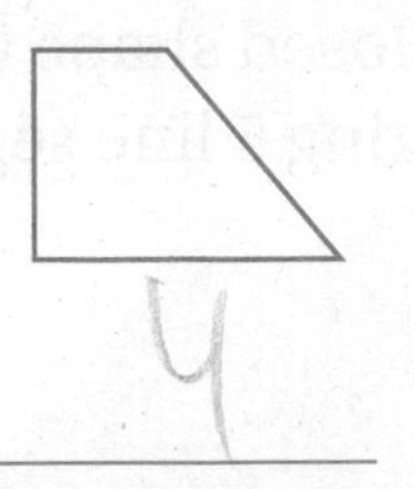

4 line segments

13. 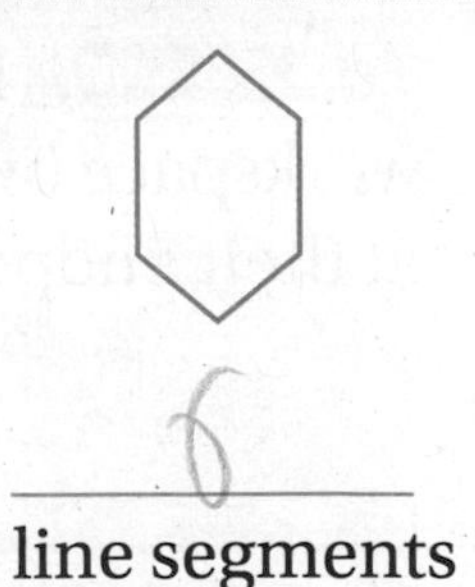

6 line segments

Write whether the shape is *open* or *closed*.

14. 

closed

15. 

open

16. 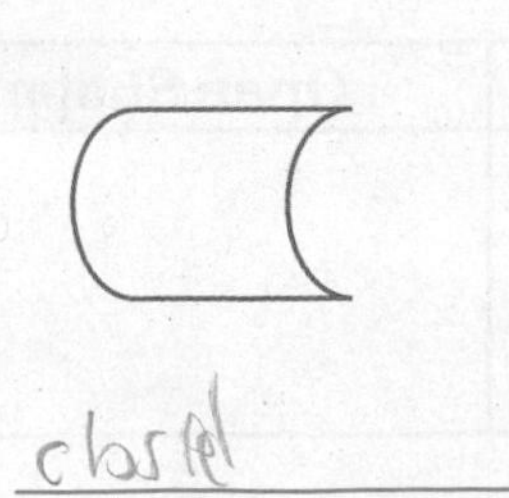

closed

17. 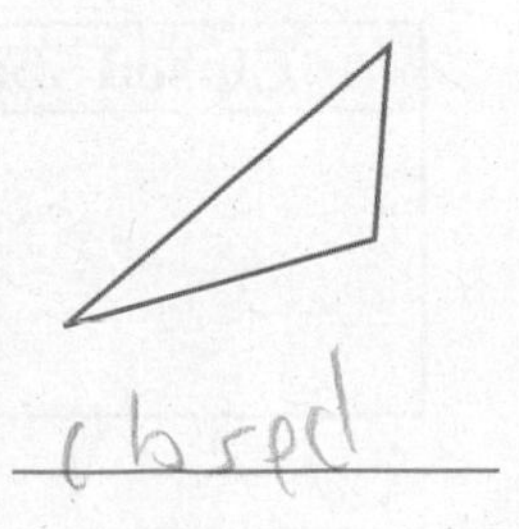

closed

Problem Solving • Applications

18. What's the Error? Brittany says there are two endpoints in the shape shown at the right. Is she correct? Explain.

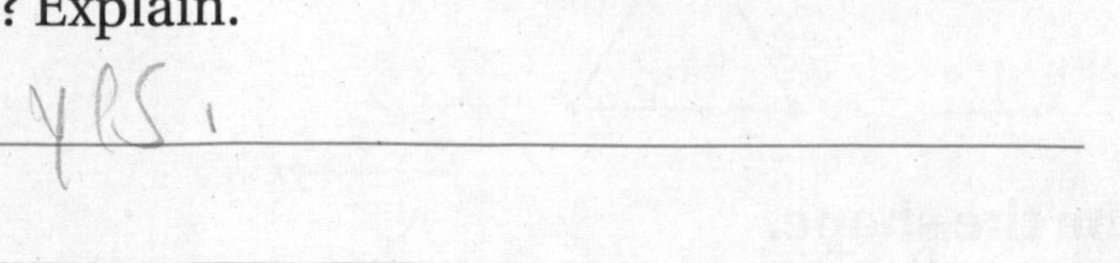

yes.

19. MATHEMATICAL PRACTICE 6 **Explain** how you can make the shape at the right a closed shape. Change the shape so it is a closed shape.

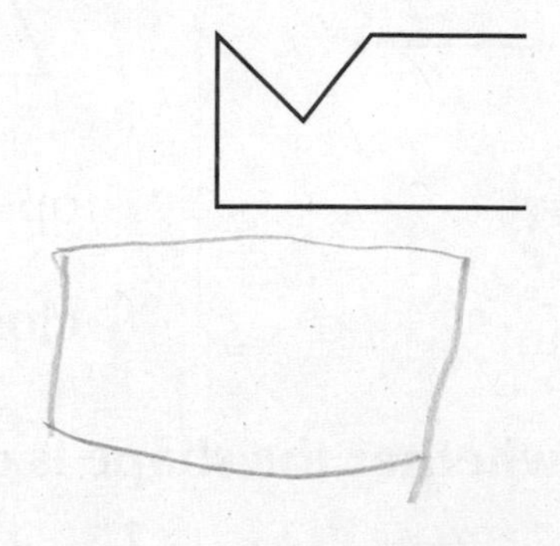

you need to fill it

20. GO DEEPER Look at Carly's drawing at the right. What did she draw? How is it like a line? How is it different? Change the drawing so that it is a line.

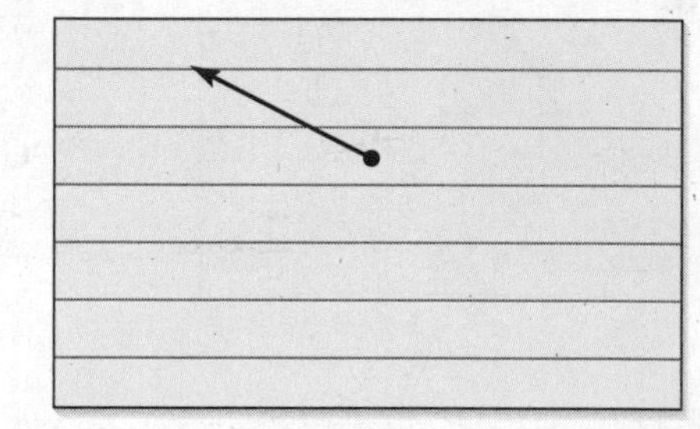

21. THINK SMARTER Draw a closed shape in the workspace by connecting 5 line segments at their endpoints.

22. THINK SMARTER Draw each shape where it belongs in the table.

Closed Shape	Open Shape

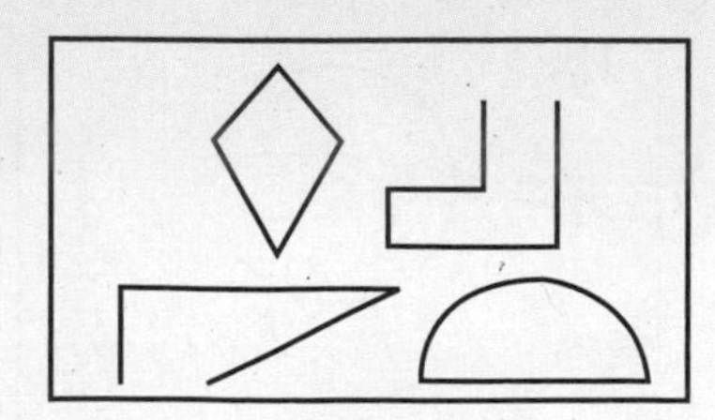

Name ____________________

Describe Angles in Plane Shapes

Essential Question How can you describe angles in plane shapes?

Common Core Geometry—3.G.A.1

MATHEMATICAL PRACTICES
MP2, MP4, MP5

Unlock the Problem

An **angle** is formed by two rays that share an endpoint. Plane shapes have angles formed by two line segments that share an endpoint. The shared endpoint is called a **vertex**. The plural of *vertex* is *vertices*.

vertex

Jason drew this shape on dot paper.

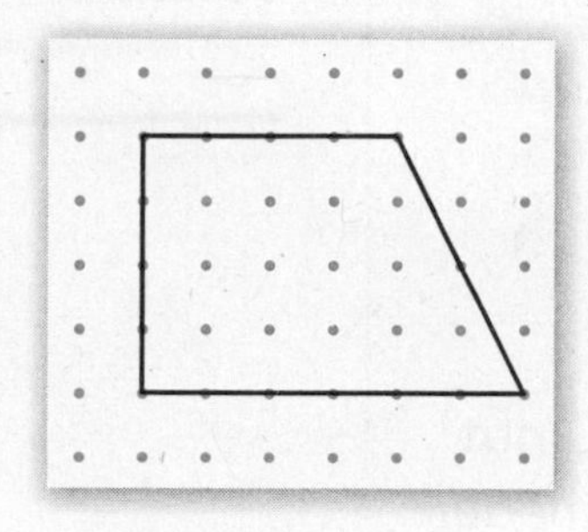

- How many angles are in Jason's shape? 4

Look at the angles in the shape that Jason drew.

How can you describe the angles?

Describe angles.

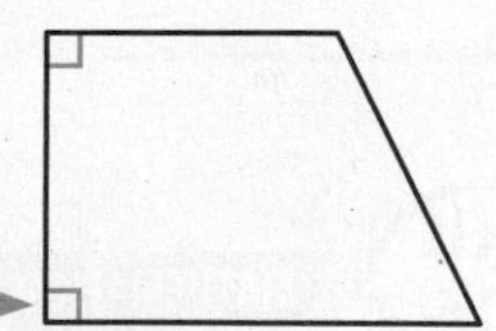

A **right angle** is an angle that forms a square corner.

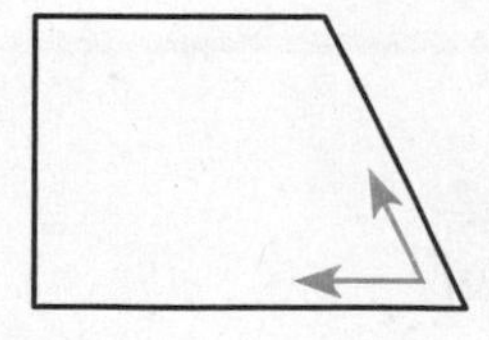

Some angles are less than a right angle.

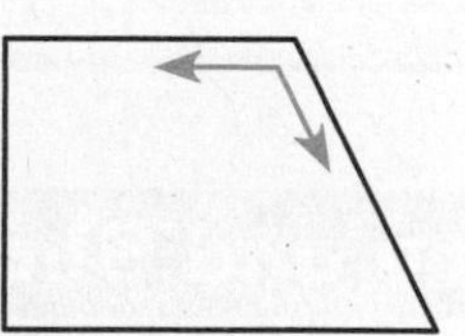

Some angles are greater than a right angle.

Look at Jason's shape.

Two angles are right angles, one angle is l than a right angle, and one angle is ggreterangle a right angle.

MATHEMATICAL PRACTICES 1

Analyze What are some examples of these types of angles that you see in everyday life? Describe where you see them and what types of angles you see.

Activity Model angles.

Materials ■ bendable straws ■ scissors ■ paper ■ pencil

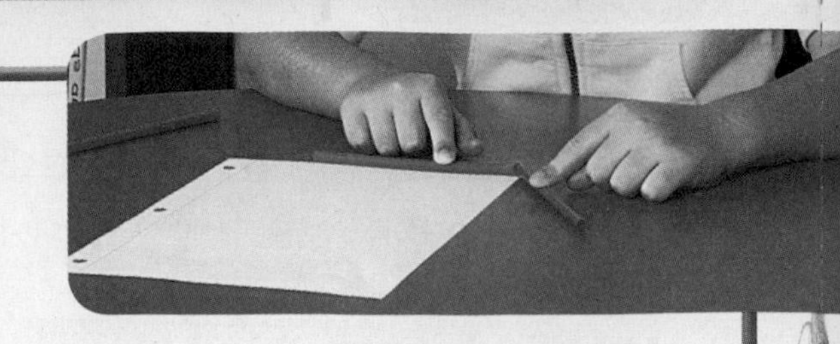

- Cut a small slit in the shorter section of a bendable straw. Cut off the shorter section of a second straw and the bendable part. Insert the slit end of the first straw into the second straw.

cut slit — cut off — insert

straw 1 — straw 2 — straw 1 — straw 2

- Make an angle with the straws you put together. Compare the angle you made to a corner of the sheet of paper.
- Open and close the straws to make other types of angles.

In the space below, trace the angles you made with the straws. Label each *right angle, less than a right angle,* or *greater than a right angle.*

Share and Show

1. How many angles are in the triangle at the right?

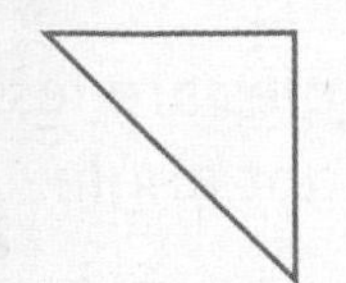

3

Math Talk MATHEMATICAL PRACTICES 2

Use Reasoning How do you know an angle is greater than or less than a right angle?

Use the corner of a sheet of paper to tell whether the angle is a *right angle, less than a right angle,* or *greater than a right angle.*

2. 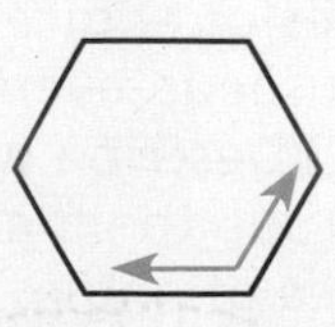

right angle

3. 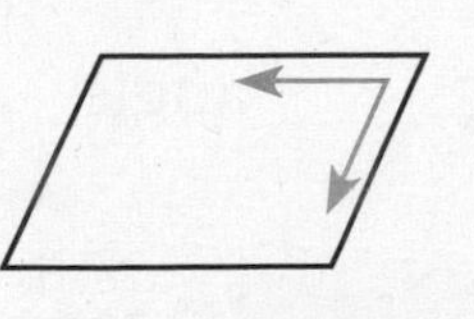

right angle

4. 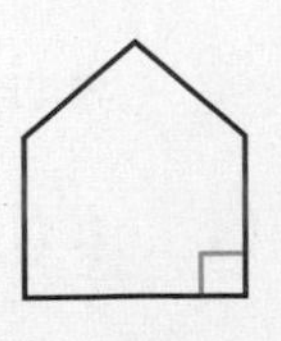

right angle

Name ______________________________

Describe Sides of Polygons

Essential Question How can you describe line segments that are sides of polygons?

Geometry—3.G.A.1

MATHEMATICAL PRACTICES
MP1, MP4, MP8

Unlock the Problem

Look at the polygon. How many pairs of sides are parallel?

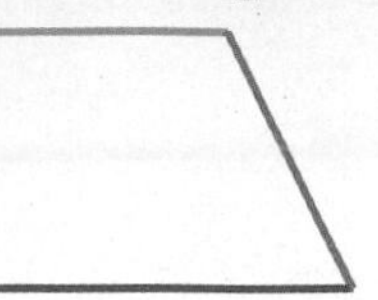

- How do you know the shape is a polygon?

____

____

TYPES OF LINES	TYPES OF LINE SEGMENTS
Lines that cross or meet are **intersecting lines**. Intersecting lines form angles. 	The orange and blue line segments meet and form an angle. So, they are intersecting.
Intersecting lines that cross or meet to form right angles are **perpendicular lines**. 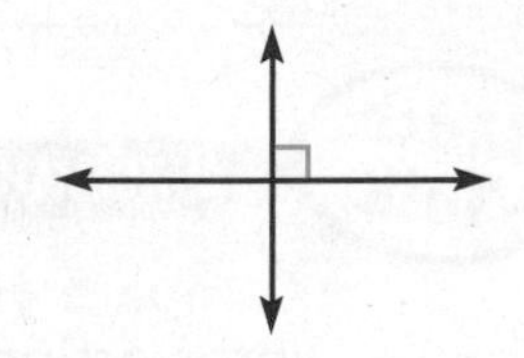	The red and blue line segments meet to form a right angle. So, they are ____________.
Lines that appear to never cross or meet and are always the same distance apart are **parallel lines**. They do not form any angles. 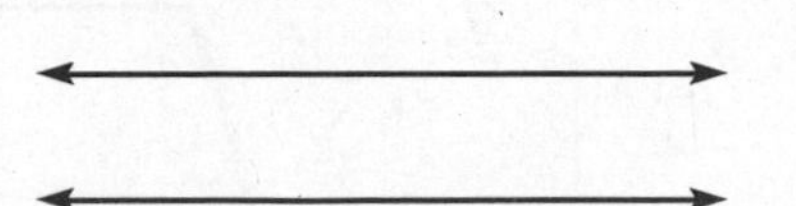	The green and blue line segments would never cross or meet. They are always the same distance apart. So, they appear to be ____________.

So, the polygon above has _____ pair of parallel sides.

Math Talk MATHEMATICAL PRACTICES 2

Use Reasoning Why can't parallel lines ever cross?

Try This! Draw a polygon with only 1 pair of parallel sides. Then draw a polygon with 2 pairs of parallel sides. Outline each pair of parallel sides with a different color.

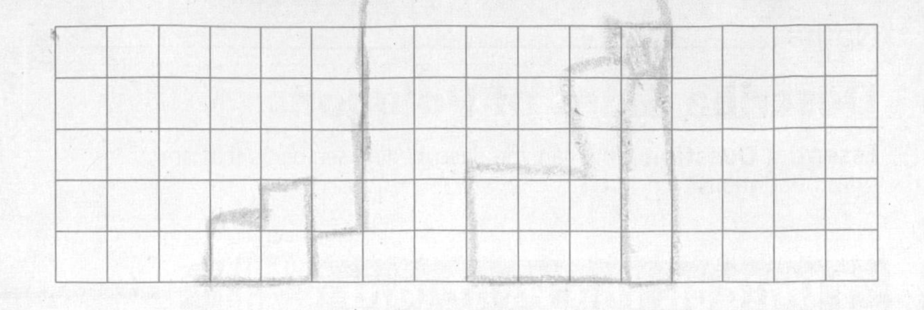

Share and Show

1. Which sides appear to be parallel?

b and C, b and d

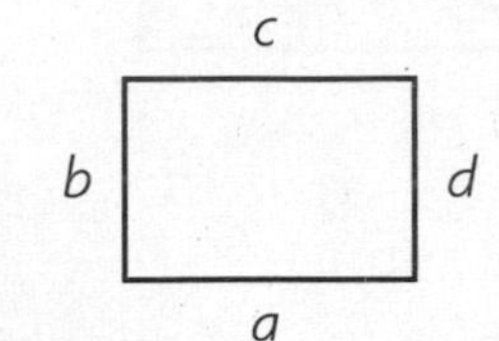

Think: Which pairs of sides appear to be the same distance apart?

Look at the green sides of the polygon. Tell if they appear to be *intersecting, perpendicular,* or *parallel.* Write all the words that describe the sides.

2.

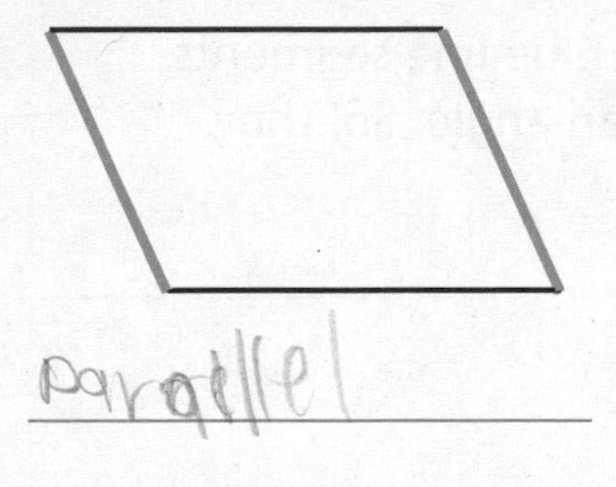

parallel

✓3.

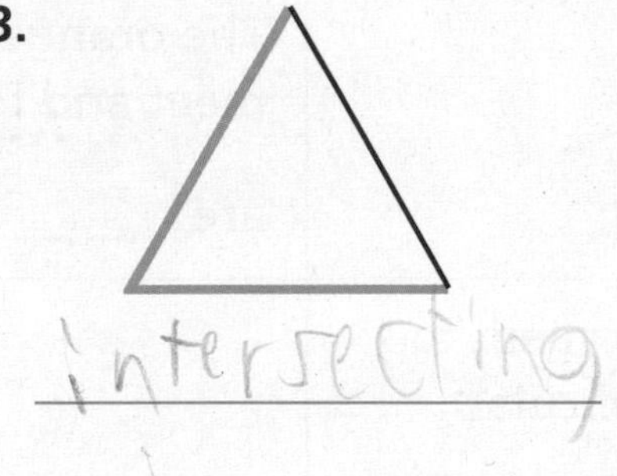

intersecting

✓4.

intersectin

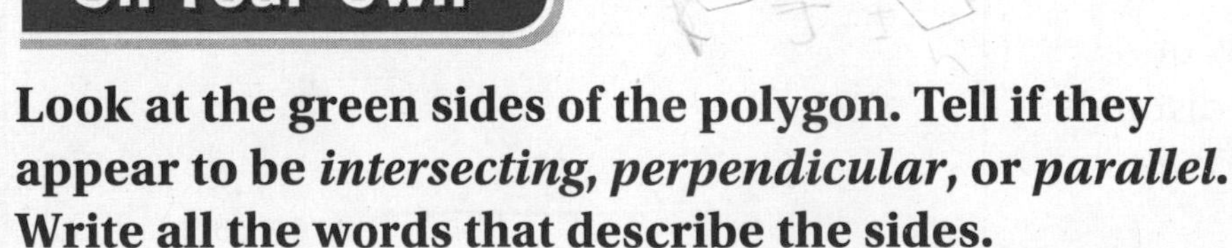

Math Talk MATHEMATICAL PRACTICES 6

Compare How are intersecting and perpendicular lines alike and how they are different?

On Your Own

Look at the green sides of the polygon. Tell if they appear to be *intersecting, perpendicular,* or *parallel.* Write all the words that describe the sides.

5.

parallel

6.

intersecting

7.

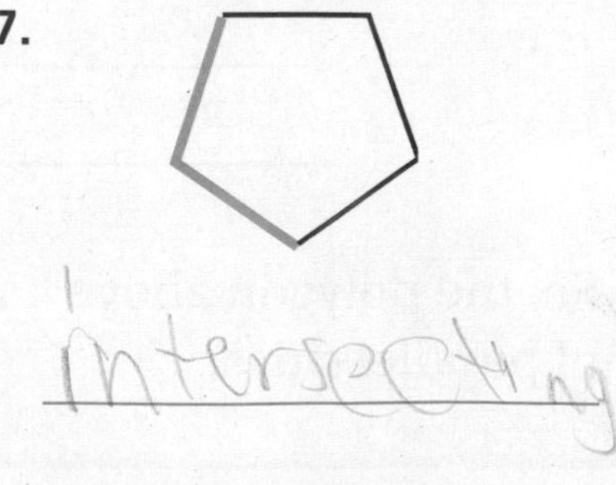

intersecting

Name ______________________________

Problem Solving • Applications

Use pattern blocks *A–E* for 8–11.

Chelsea wants to sort pattern blocks by the types of sides.

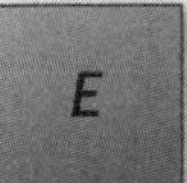

8. Which blocks have intersecting sides?

A and d

9. Which blocks have parallel sides?

E and C, A, D, E

10. Which blocks have perpendicular sides?

____

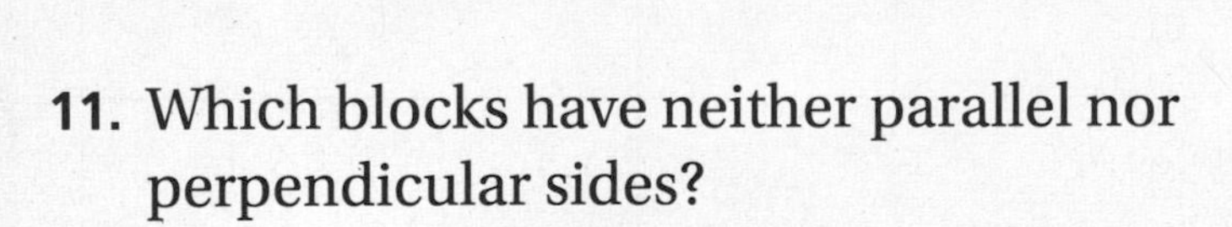

11. Which blocks have neither parallel nor perpendicular sides?

____

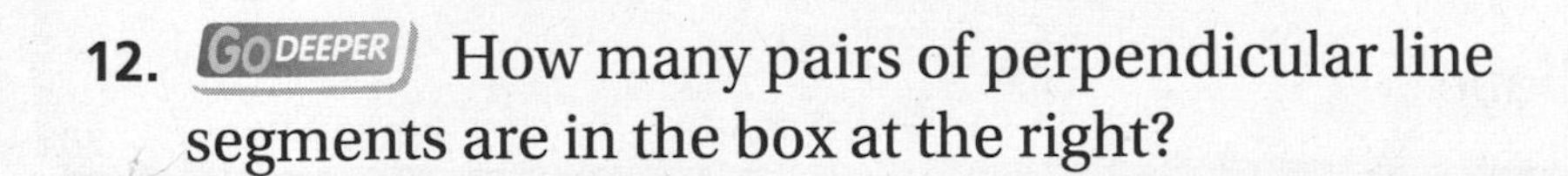

12. GO DEEPER How many pairs of perpendicular line segments are in the box at the right?

____

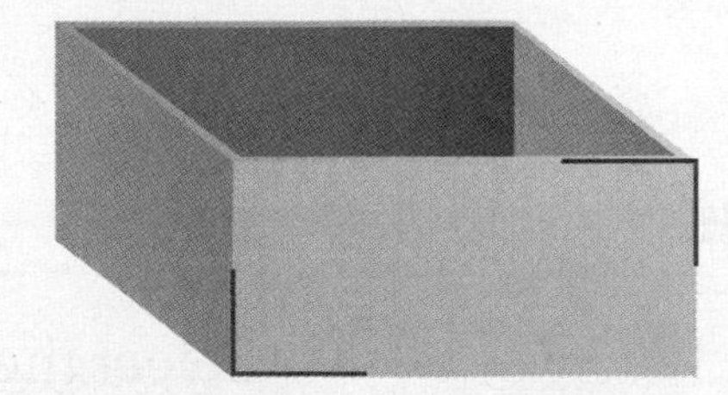

▲ The red line segments show 1 pair of perpendicular line segments.

13. THINK SMARTER Can the same two lines be parallel, perpendicular, and intersecting at the same time? Explain your answer.

____

____

____

Unlock the Problem

14. MATHEMATICAL PRACTICE 3 **Compare Representations** I am a pattern block that has 2 fewer sides than a hexagon. I have 2 pairs of parallel sides and 4 right angles. Which shape am I?

a. What do you need to know? ______________________

____

b. How can you find the answer to the riddle? ______________________

____

c. Write *yes* or *no* in the table to solve the riddle.

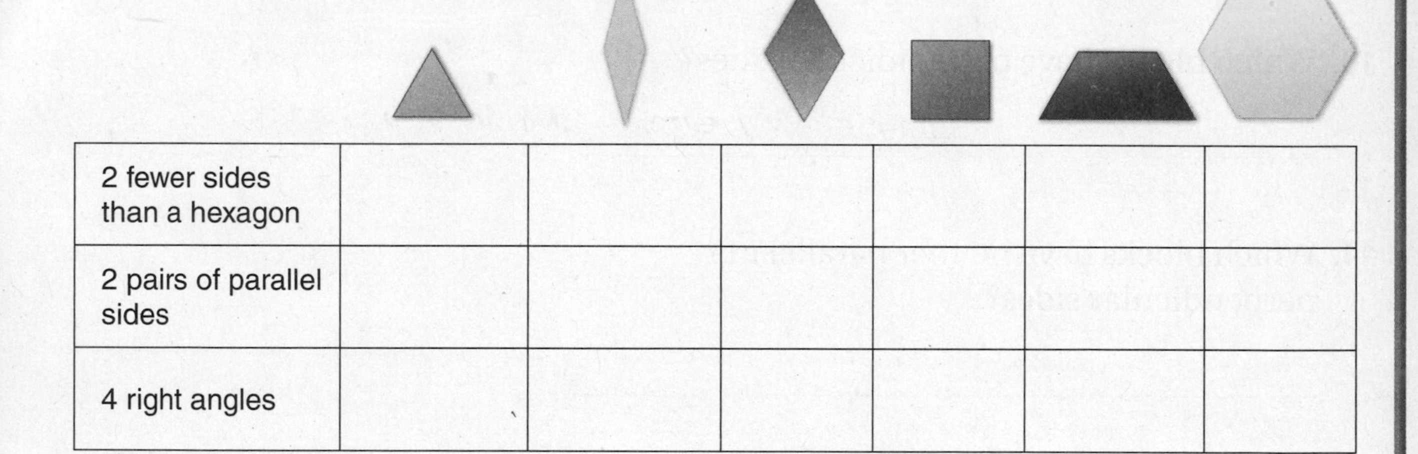

2 fewer sides than a hexagon							
2 pairs of parallel sides							
4 right angles							

So, the ______________ is the shape.

15. THINK SMARTER Select the shapes that have at least one pair of parallel sides. Mark all that apply.

 A 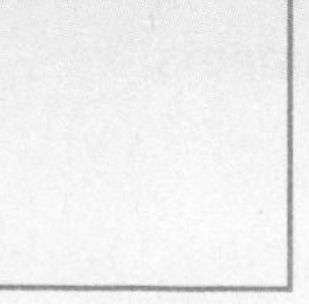

 C 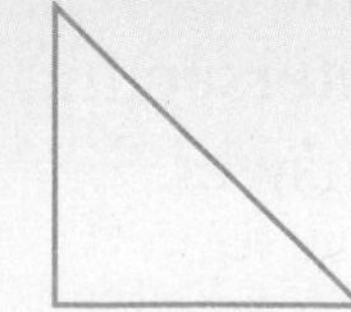

B 

D 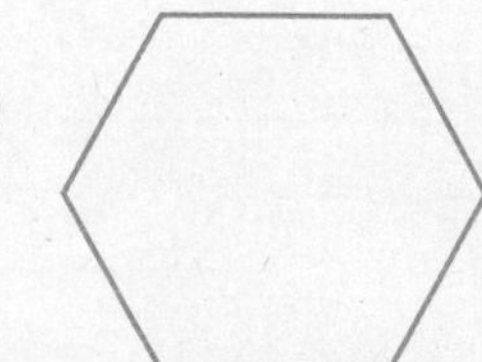

Name ______________________________

Mid-Chapter Checkpoint

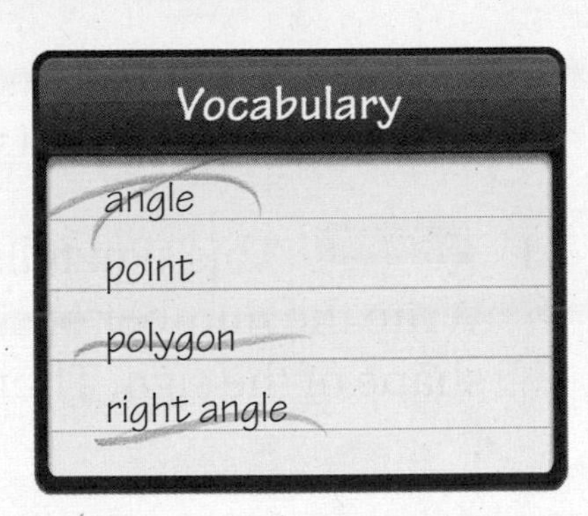

Vocabulary

Choose the best term from the box to complete the sentence.

1. An __angle__ is formed by two rays that share an endpoint. (p. 703)
2. A __polygon__ is a closed shape made up of line segments. (p. 709)
3. A __right angle__ forms a square corner. (p. 703)

Concepts and Skills

Use the corner of a sheet of paper to tell whether the angle is a *right angle, less than a right angle,* or *greater than a right angle.* (3.G.A.1)

4. 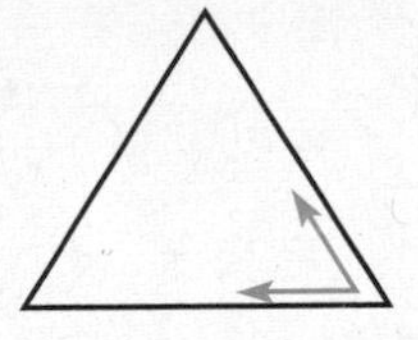

__less than a right__

5. 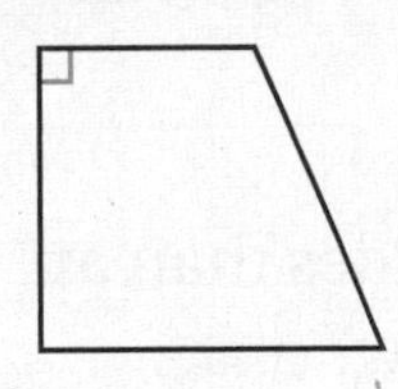

__right angle__

6. 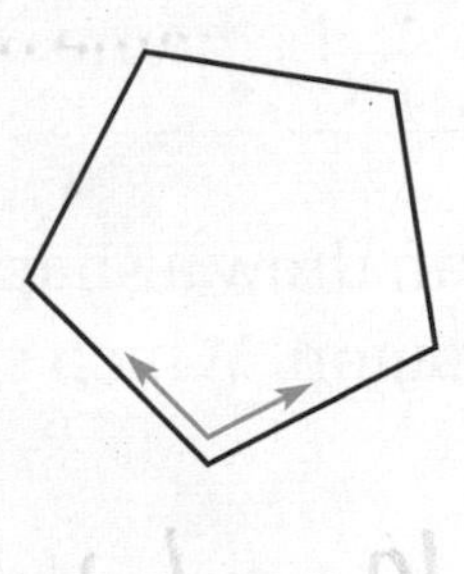

__right angle__

Write the number of sides and the number of angles. Then name the polygon. (3.G.A.1)

7. 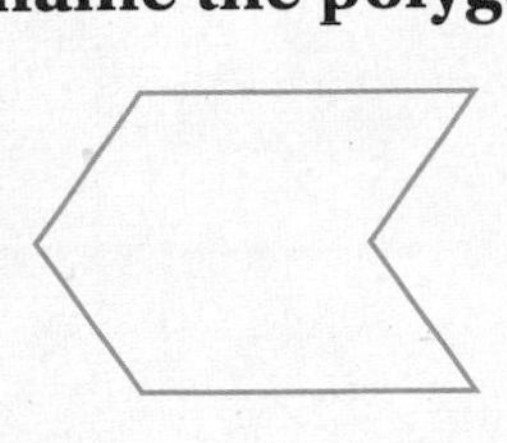

__6__ sides

__6__ angles

__hexagon__

8. 

__4__ sides

__4__ angles

__quadrilateral__

9. 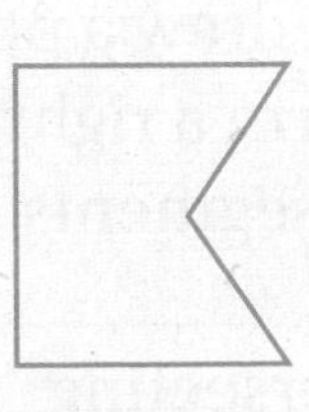

__5__ sides

__5__ angles

__pentagon__

10. Anne drew the shape at the right. Is her shape an open shape or a closed shape? (3.G.A.1)

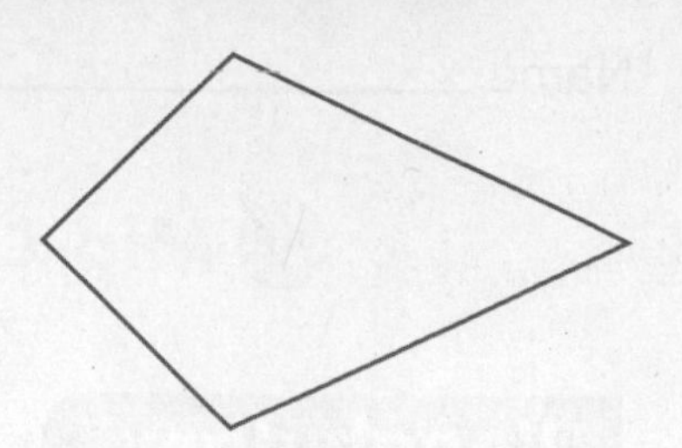

closed shape

11. GO DEEPER This sign tells drivers there is a steep hill ahead. Write the number of sides and the number of angles in the shape of the sign. Then name the shape. (3.G.A.1)

4 angles 4 sides quadrilateral

12. Why is this closed plane shape NOT a polygon? (3.G.A.1)

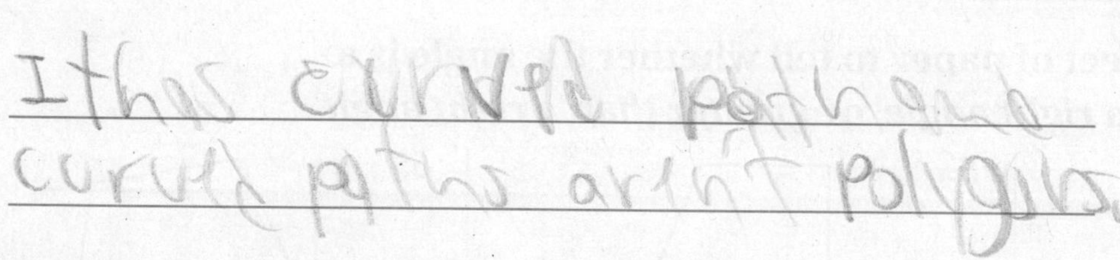

13. Sean drew a shape with 2 fewer sides than an octagon. Which shape did he draw? (3.G.A.1)

14. John drew a polygon with two line segments that meet to form a right angle. Circle the words that describe the line segments. (3.G.A.1)

intersecting
curved
parallel
perpendicular

Name ______________________

Classify Quadrilaterals

Geometry—3.G.A.1

MATHEMATICAL PRACTICES
MP2, MP4, MP6

Essential Question How can you use sides and angles to help you describe quadrilaterals?

Unlock the Problem

Quadrilaterals are named by their sides and their angles.

Describe quadrilaterals.

quadrilateral

_______ sides

_______ angles

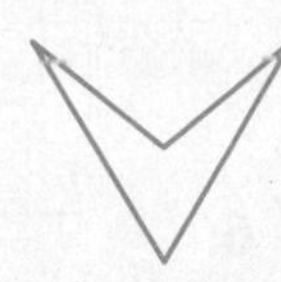

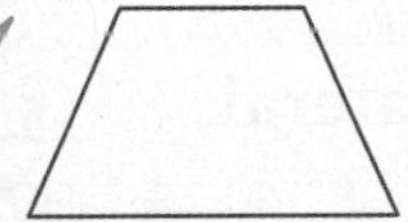

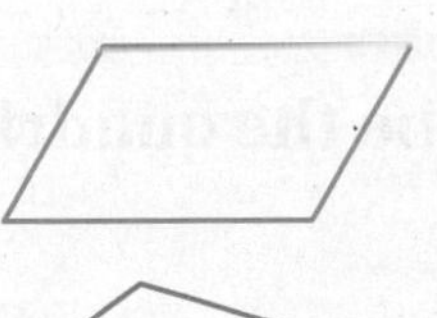

ERROR Alert

Some quadrilaterals cannot be classified as a trapezoid, rectangle, square, or rhombus.

trapezoid

exactly _______ pair of opposite sides that are parallel

lengths of sides could be the same

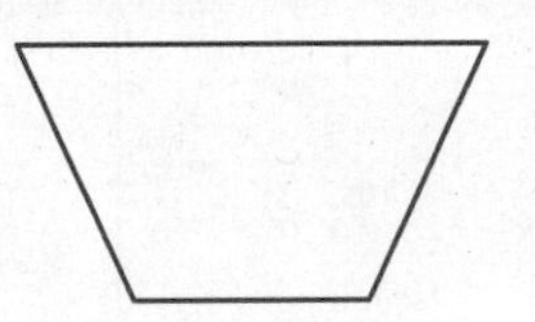

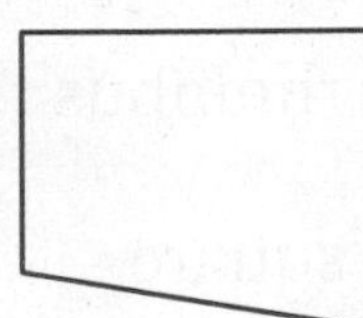

rectangle

_____ pairs of opposite sides that are parallel

_____ pairs of sides that are of equal length

_____ right angles

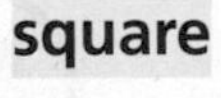

square

_____ pairs of opposite sides that are parallel

_____ sides that are of equal length

_____ right angles

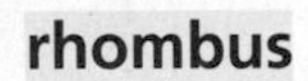

rhombus

_____ pairs of opposite sides that are parallel

_____ sides that are of equal length

Math Talk MATHEMATICAL PRACTICES 8

Generalize Why can a square also be named a rectangle or a rhombus?

Share and Show 

Look at the quadrilateral at the right.

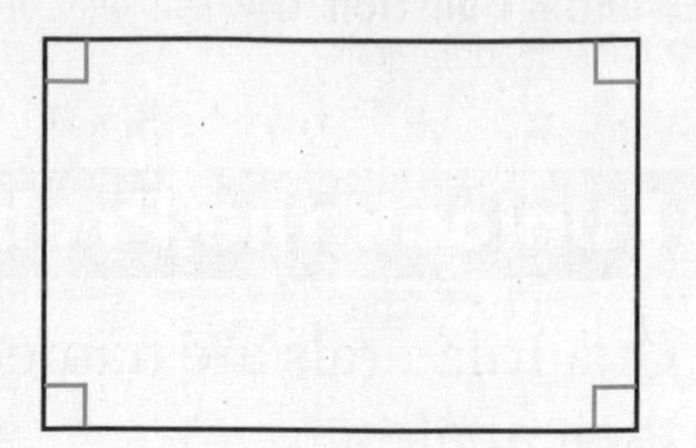

Think: All the angles are right angles.

1. Outline each pair of opposite sides that are parallel with a different color. How many pairs of opposite sides appear to be parallel? ________

2. Look at the parallel sides you colored.

   The sides in each pair are of ________ length.

3. Name the quadrilateral. ________________

Circle all the words that describe the quadrilateral.

4. 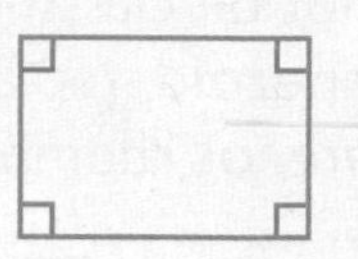

   rectangle

   rhombus

   square

   trapezoid

5. 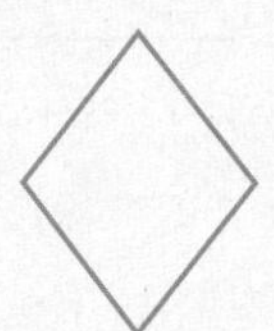

   rhombus

   quadrilateral

   square

   rectangle

6. 

   rectangle

   rhombus

   trapezoid

   quadrilateral

Math Talk MATHEMATICAL PRACTICES 1

Analyze How can you have a rhombus that is not a square?

On Your Own

Circle all the words that describe the quadrilateral.

7. 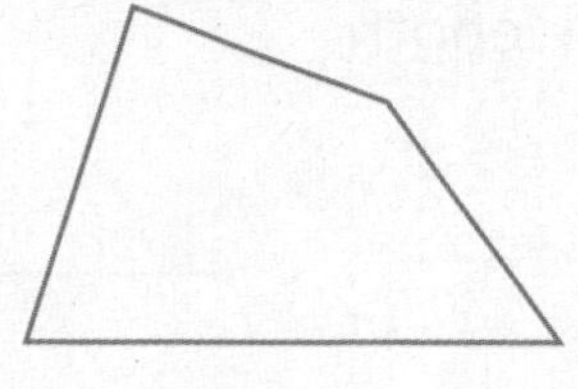

   rectangle

   trapezoid

   quadrilateral

   rhombus

8. 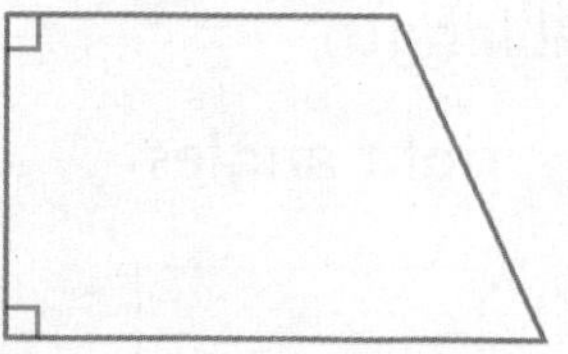

   rectangle

   rhombus

   trapezoid

   square

9. 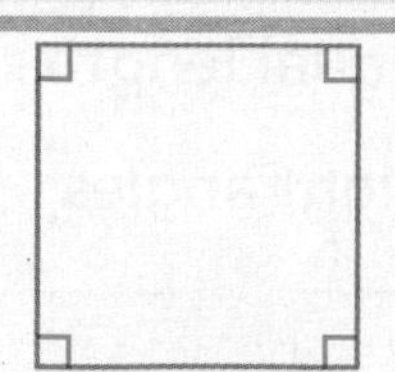

   quadrilateral

   square

   rectangle

   rhombus

Name ____________________

Problem Solving • Applications

Use the quadrilaterals at the right for 10–12.

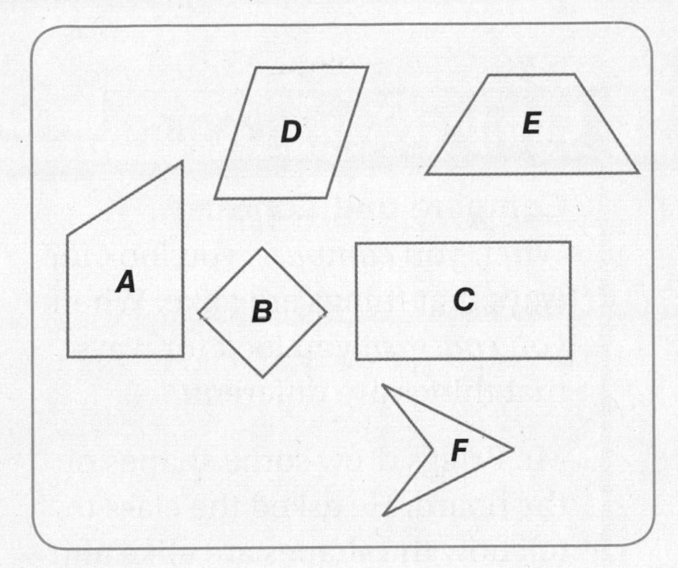

10. Which quadrilaterals appear to have 4 right angles?

____

11. Which quadrilaterals appear to have 2 pairs of opposite sides that are parallel?

____

12. Which quadrilaterals appear to have no right angles?

____

Write *all* or *some* to complete the sentence for 13–18.

13. The opposite sides of _____ rectangles are parallel.

14. _____ sides of a rhombus are the same length.

15. _____ squares are rectangles.

16. _____ rhombuses are squares.

17. _____ quadrilaterals are polygons.

18. _____ polygons are quadrilaterals.

19. MATHEMATICAL PRACTICE 6 Circle the shape at the right that is not a quadrilateral. **Explain** your choice.

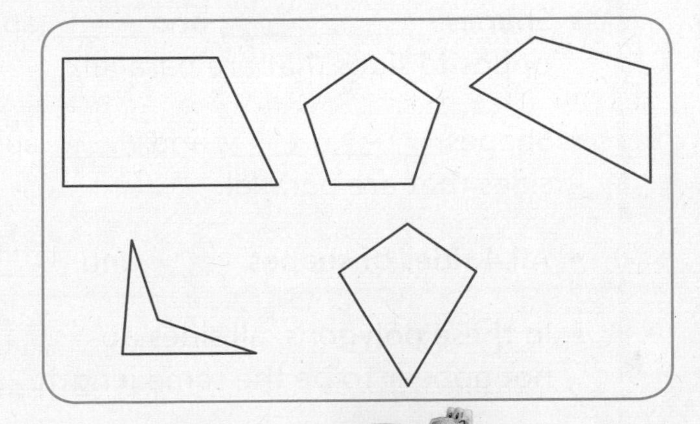

____

____

____

20. THINKSMARTER I am a polygon that has 4 sides and 4 angles. At least one of my angles is less than a right angle. Circle all the shapes that I could be.

quadrilateral   rectangle   square   rhombus   trapezoid

21. THINKSMARTER Identify the quadrilateral that can have two pairs of parallel sides and no right angles.

Ⓐ rhombus Ⓑ square Ⓒ trapezoid

Connect to Reading

Compare and Contrast

When you *compare*, you look for ways that things are alike. When you *contrast*, you look for ways that things are different.

Mr. Briggs drew some shapes on the board. He asked the class to tell how the shapes are alike and how they are different.

GO DEEPER **Complete the sentences.**

- Shapes _____, _____, _____, _____, _____, _____, and _____ are polygons.
- Shapes _____, _____, and _____ are not polygons.
- Shapes _____, _____, _____, and _____ are quadrilaterals.
- Shapes _____, _____, and _____ appear to have only 1 pair of opposite sides that are parallel.
- Shapes _____, _____, and _____ appear to have 2 pairs of opposite sides that are parallel.
- All 4 sides of shapes _____ and _____ appear to be the same length.
- In these polygons, all sides do not appear to be the same length. _______________
- These shapes can be called rhombuses. _______________
- Shapes _____ and _____ are quadrilaterals, but cannot be called rhombuses.
- Shape _____ is a rhombus and can be called a square.

Name ___________________________

Classify Quadrilaterals

Common Core **COMMON CORE STANDARD—3.G.A.1**
Reason with shapes and their attributes.

Circle all the words that describe the quadrilateral.

1. 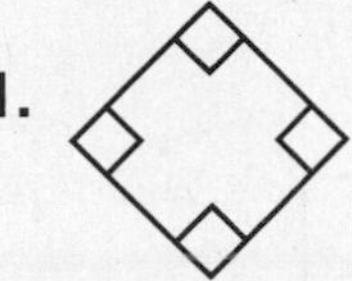

square

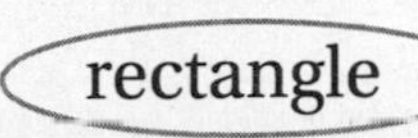

rectangle

rhombus

trapezoid

2.

square

rectangle

rhombus

trapezoid

3.

square

rectangle

rhombus

trapezoid

Use the quadrilaterals below for 4–6.

A

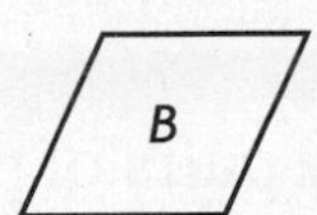

C

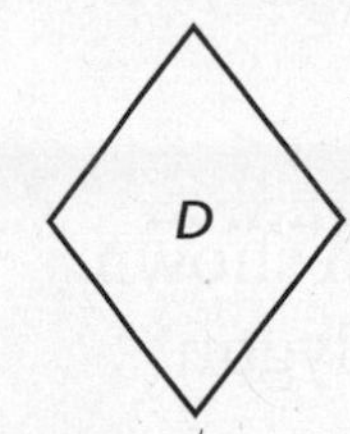

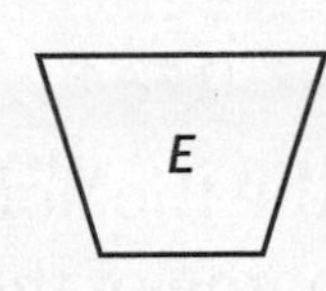

4. Which quadrilaterals appear to have no right angles?

____

5. Which quadrilaterals appear to have 4 right angles?

____

6. Which quadrilaterals appear to have 4 sides of equal length?

____

Problem Solving Real World

7. A picture on the wall in Jeremy's classroom has 4 right angles, 4 sides of equal length, and 2 pairs of opposite sides that are parallel. What quadrilateral best describes the picture?

____

8. WRITE ▸ *Math* Explain how a trapezoid and a rectangle are different.

____

____

____

____

Lesson Check (3.G.A.1)

1. What word describes the quadrilateral?

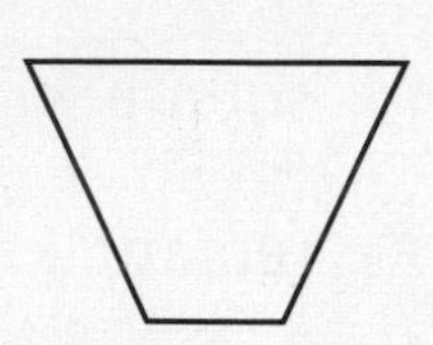

2. Which quadrilaterals appear to have 2 pairs of opposite sides that are parallel?

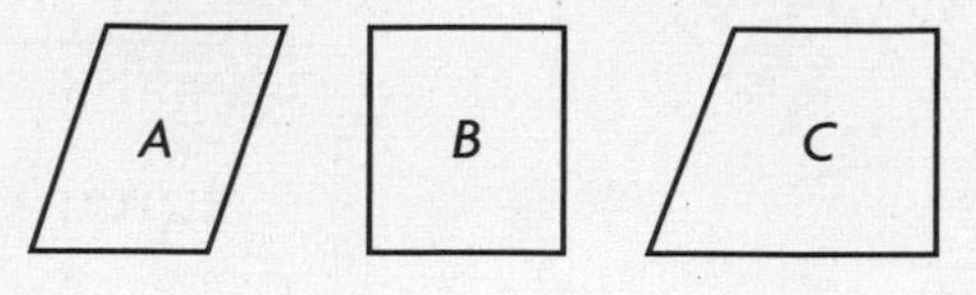

Spiral Review (3.G.A.1)

3. Aiden drew the the polygon shown. What is the name of the polygon he drew?

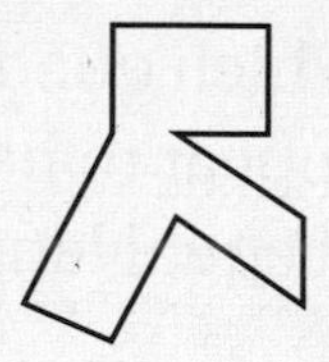

4. How many pairs of parallel sides does this shape appear to have?

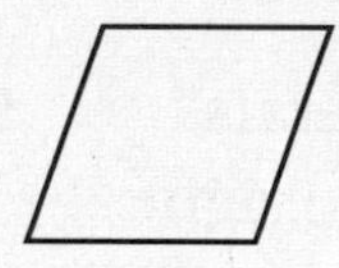

5. What word describes the dashed sides of the shape shown?

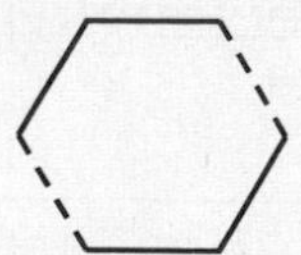

6. How many right angles does this shape have?

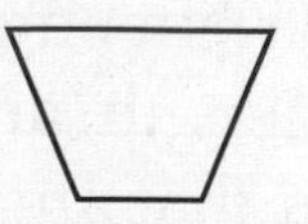

FOR MORE PRACTICE GO TO THE **Personal Math Trainer**

Name ____________________________

Draw Quadrilaterals

Essential Question How can you draw quadrilaterals?

Geometry—3.G.A.1

MATHEMATICAL PRACTICES
MP3, MP6, MP7, MP8

Unlock the Problem

Hands On

CONNECT You have learned to classify quadrilaterals by the number of pairs of opposite sides that are parallel, by the number of pairs of sides of equal length, and by the number of right angles.

How can you draw quadrilaterals?

Activity 1 Use grid paper to draw quadrilaterals.

Materials ■ ruler

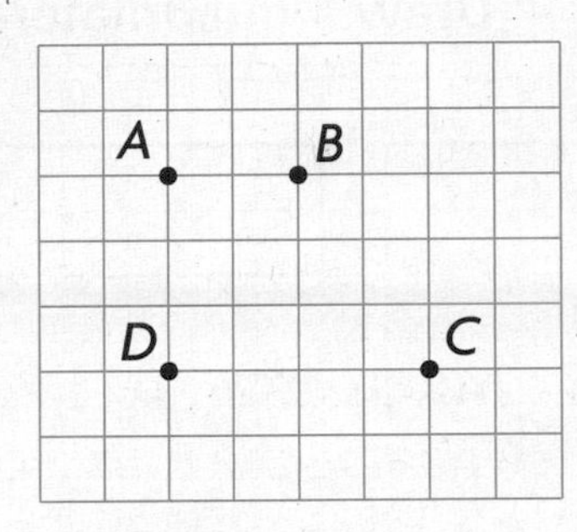

- Use a ruler to draw line segments from points A to B, from B to C, from C to D, and from D to A.
- Write the name of your quadrilateral.

____

Activity 2 Draw a shape that does not belong.

Materials ■ ruler

A **Here are three examples of a quadrilateral. Draw an example of a polygon that is not a quadrilateral.**

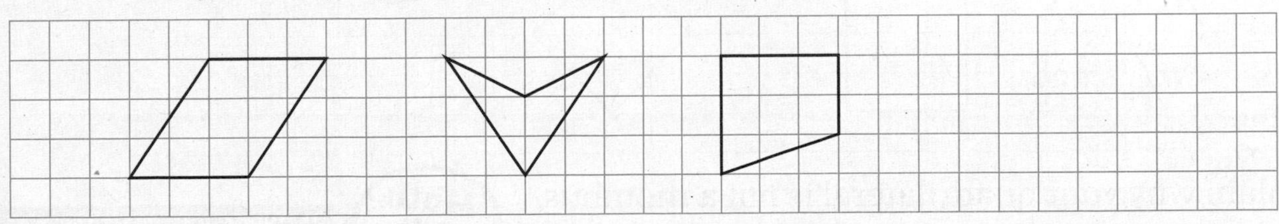

- Explain why your polygon is not a quadrilateral.

____

____

B **Here are three examples of a square.**
Draw a quadrilateral that is not a square.

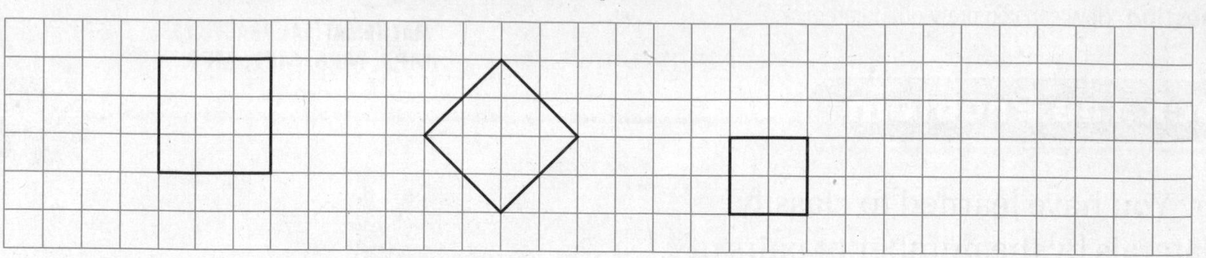

- Explain why your quadrilateral is not a square.

____

____

C **Here are three examples of a rectangle.**
Draw a quadrilateral that is not a rectangle.

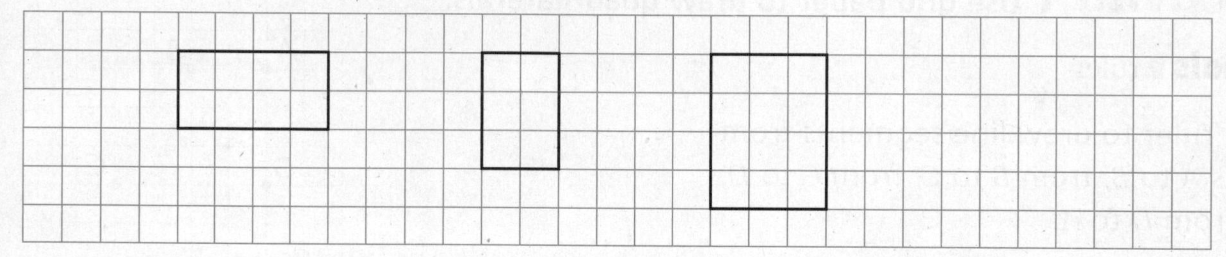

- Explain why your quadrilateral is not a rectangle.

____

____

D **Here are three examples of a rhombus.**
Draw a quadrilateral that is not a rhombus.

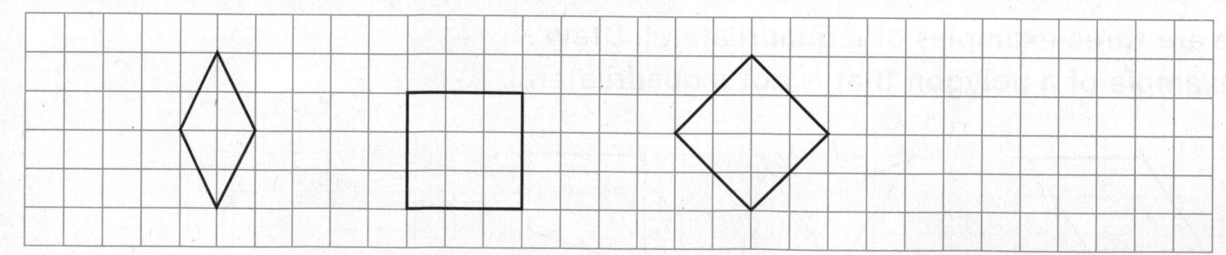

- Explain why your quadrilateral is not a rhombus.

____

____

MATHEMATICAL PRACTICES 3

Compare Representations Compare your drawings with your classmates. Explain how your drawings are alike and how they are different.

Name ____________________

Share and Show

1. Choose four endpoints that connect to make a rectangle.

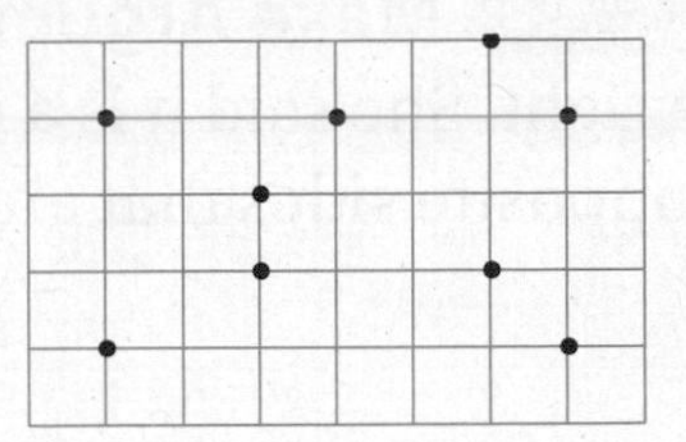

Think: A rectangle has 2 pairs of opposite sides that are parallel, 2 pairs of sides of equal length, and 4 right angles.

Draw a quadrilateral that is described. Name the quadrilateral you drew.

2. 2 pairs of equal sides

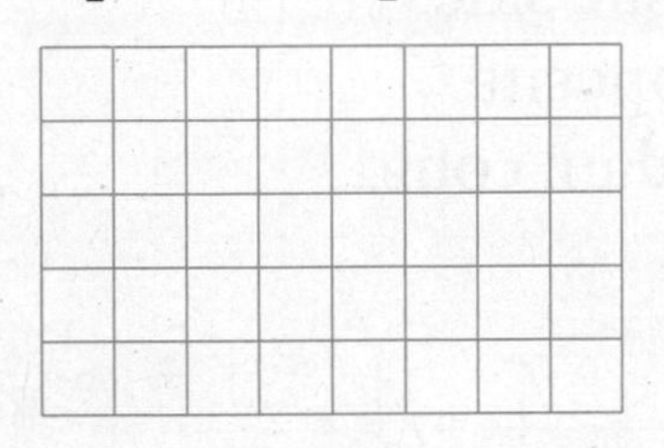

Name __________

3. 4 sides of equal length

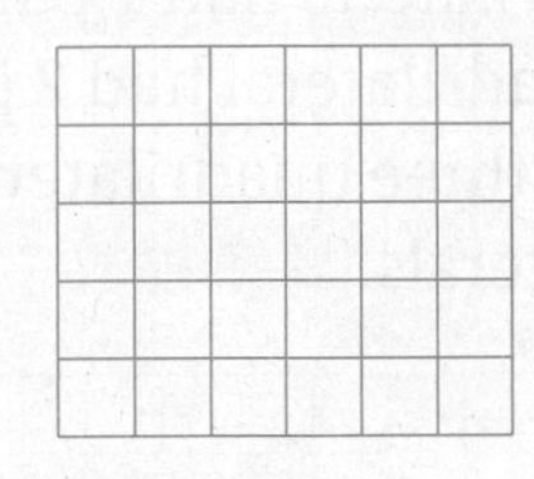

Name __________

Math Talk — MATHEMATICAL PRACTICES 6

Compare Explain one way the quadrilaterals you drew are alike and one way they are different.

On Your Own

Practice: Copy and Solve **Use grid paper to draw a quadrilateral that is described. Name the quadrilateral you drew.**

4. exactly 1 pair of opposite sides that are parallel

5. 4 right angles

6. 2 pairs of sides of equal length

Draw a quadrilateral that does not belong. Then explain why.

7.

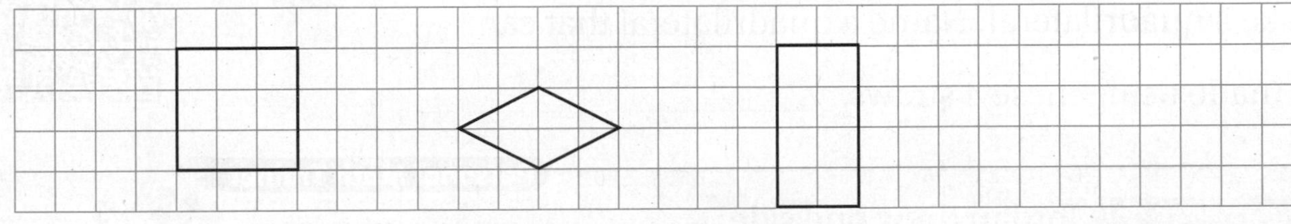

____

8.

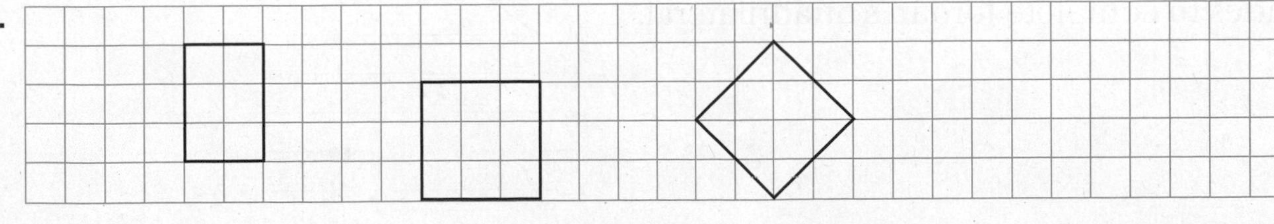

____

Problem Solving • Applications

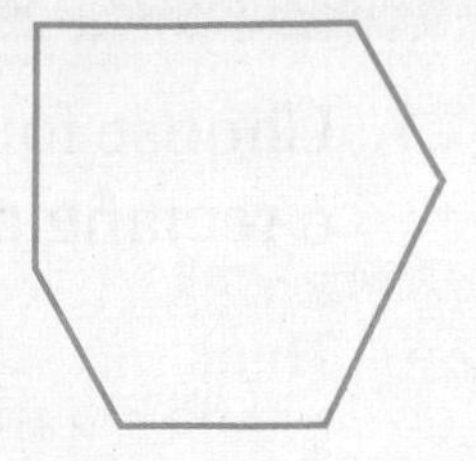

9. MATHEMATICAL PRACTICE 3 **Make Arguments** Jacki drew the shape at the right. She said it is a rectangle because it has 2 pairs of opposite sides that are parallel. Describe her error.

____

____

10. GO DEEPER Adam drew three quadrilaterals. One quadrilateral had no pairs of parallel sides, one quadrilateral had 1 pair of opposite sides that are parallel, and the last quadrilateral had 2 pairs of opposite sides that are parallel. Draw the three quadrilaterals that Adam could have drawn. Name the quadrilaterals.

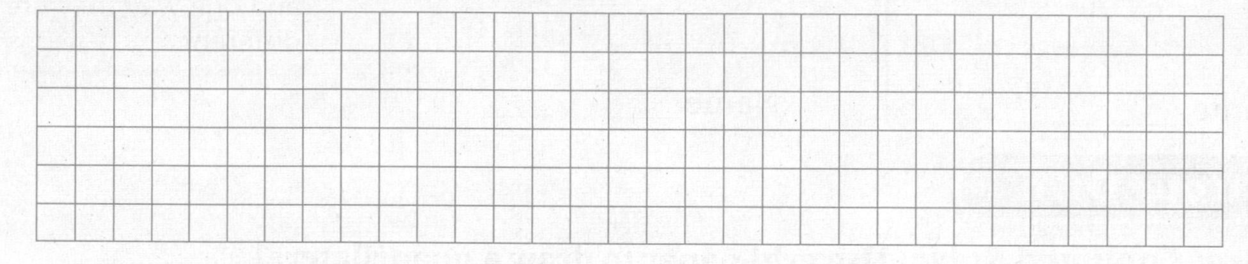

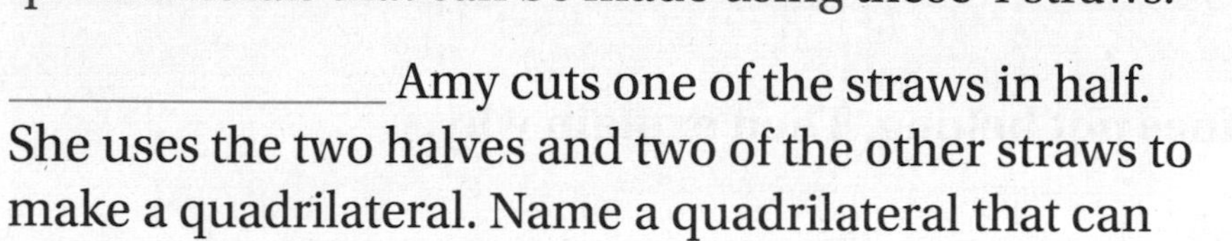

11. THINK SMARTER Amy has 4 straws of equal length. Name the quadrilaterals that can be made using these 4 straws.

_______________ Amy cuts one of the straws in half. She uses the two halves and two of the other straws to make a quadrilateral. Name a quadrilateral that can be made using these 4 straws. _______________

Personal Math Trainer

12. THINK SMARTER + Jordan drew one side of a quadrilateral with 2 pairs of opposite sides that are parallel. Draw the other 3 sides to complete Jordan's quadrilateral.

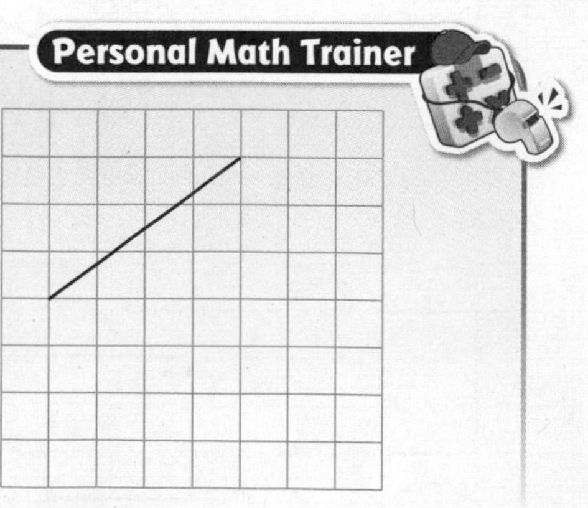

Name ______________________

Draw Quadrilaterals

Common Core **COMMON CORE STANDARD—3.G.A.1**
Reason with shapes and their attributes.

Draw a quadrilateral that is described.
Name the quadrilateral you drew.

1. 4 sides of equal length

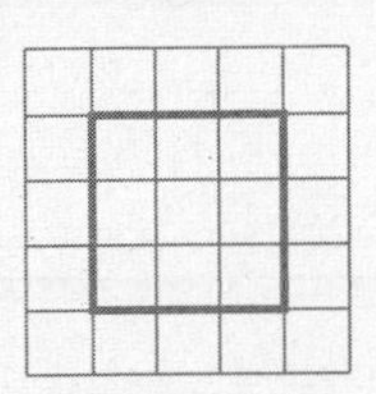

square

2. 1 pair of opposite sides that are parallel

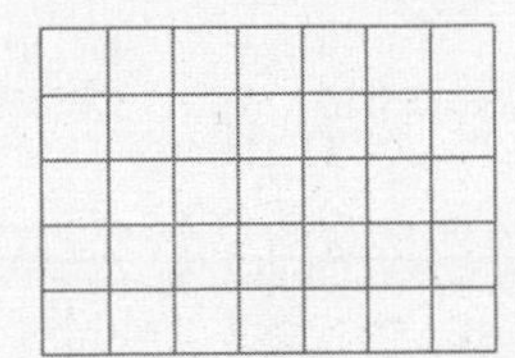

____

Draw a quadrilateral that does not belong.
Then explain why.

3.

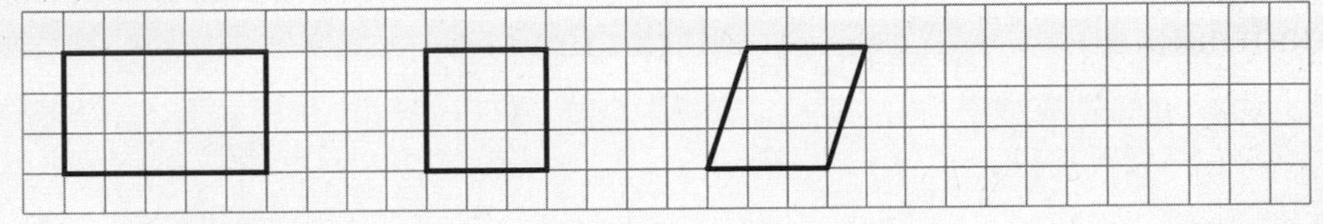

____

____

____

Problem Solving Real World

4. Layla drew a quadrilateral with 4 right angles and 2 pairs of opposite sides that are parallel. Name the quadrilateral she could have drawn.

____

5. WRITE ▸ *Math* Draw a quadrilateral that is NOT a rectangle. Describe your shape, and explain why it is not a rectangle.

____

____

Lesson Check (3.G.A.1)

1. Chloe drew a quadrilateral with 2 pairs of opposite sides that are parallel. Name a shape that could be Chloe's quadrilateral?

2. Mike drew a quadrilateral with four right angles. What shape could he have drawn?

Spiral Review (3.MD.C.7, 3.MD.D.8, 3.G.A.1)

3. A quadrilateral has 4 right angles and 4 sides of equal length. What is the name of the quadrilateral?

4. Mark drew two lines that form a right angle. What word describes the lines Mark drew?

5. Dennis drew the rectangle on grid paper. What is the perimeter of the rectangle Dennis drew?

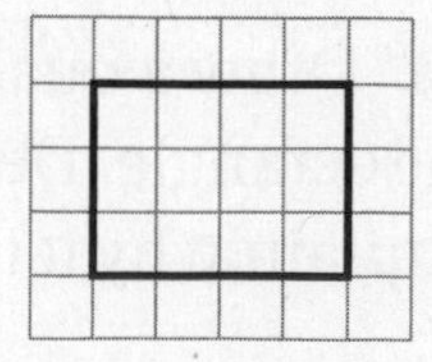

6. Jill drew the rectangle on grid paper. What is the area of the rectangle Jill drew?

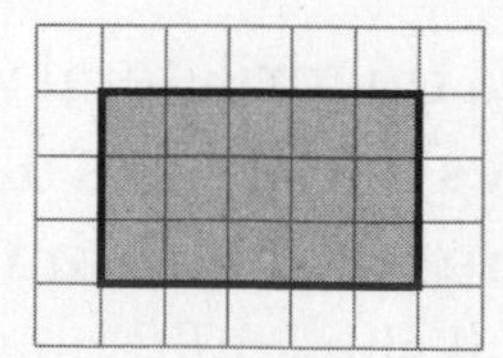

FOR MORE PRACTICE GO TO THE **Personal Math Trainer**

Name ______________________

Describe Triangles

Essential Question How can you use sides and angles to help you describe triangles?

Common Core **Geometry—3.G.A.1**

MATHEMATICAL PRACTICES MP4, MP5, MP7, MP8

Unlock the Problem

Real World

Hands On

How can you use straws of different lengths to make triangles?

Activity

Materials ■ straws ■ scissors ■ MathBoard

STEP 1 Cut straws into different lengths.

STEP 2 Find straw pieces that you can put together to make a triangle. Draw your triangle on the MathBoard.

STEP 3 Find straw pieces that you cannot put together to make a triangle.

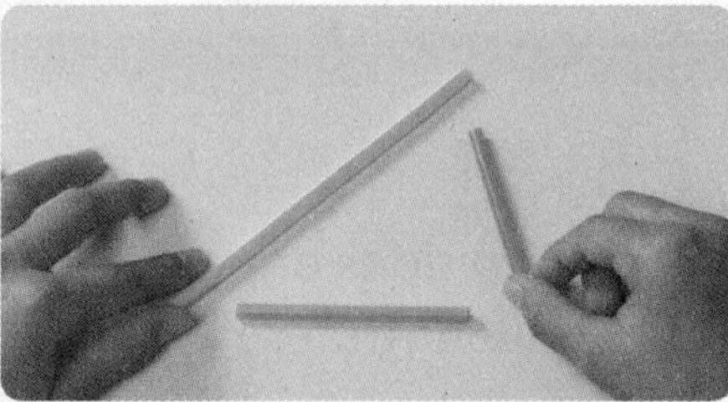

1. Compare the lengths of the sides. Describe when you can make a triangle.

____

____

Math Talk MATHEMATICAL PRACTICES 2

Reason Abstractly
What if you had three straws of equal length? Can you make a triangle?

2. MATHEMATICAL PRACTICE 1 **Describe** when you cannot make a triangle.

____

____

____

3. Explain how you can change the straw pieces in Step 3 to make a triangle. ____________

____

Ways to Describe Triangles

What are two ways triangles can be described?

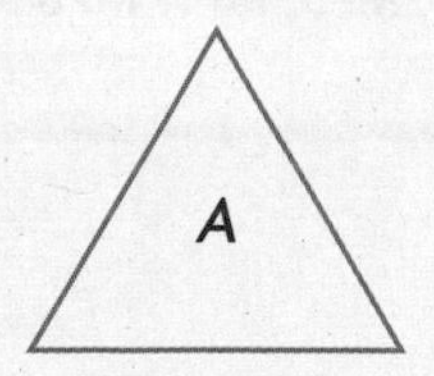

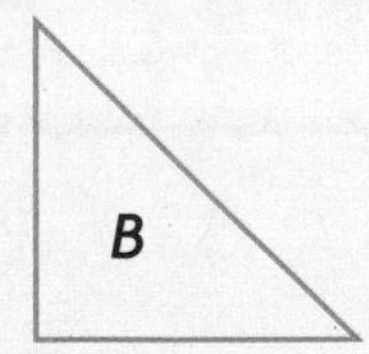

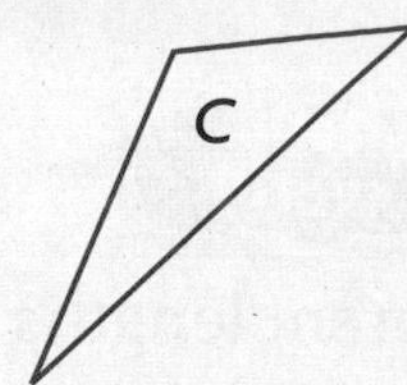

One Way

Triangles can be described by the number of sides that are of equal length.

Draw a line to match the description of the triangle(s).

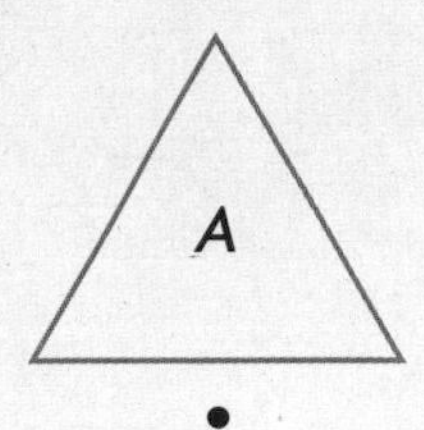

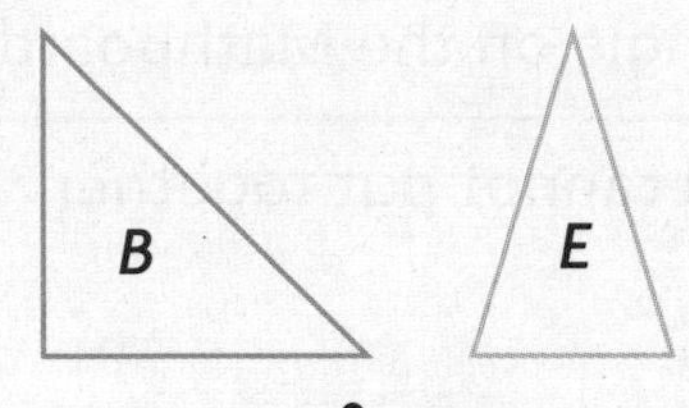

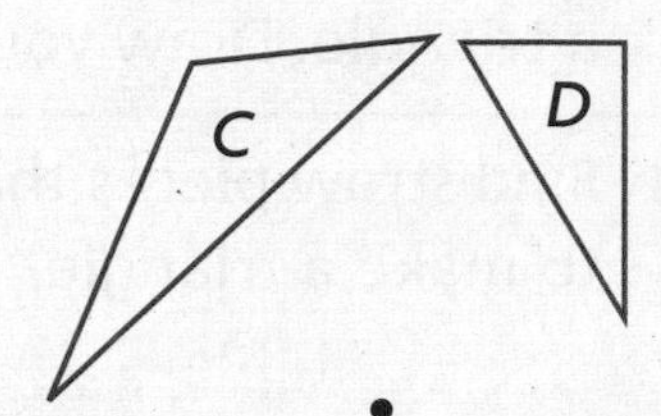

No sides are equal in length.

Two sides are equal in length.

Three sides are equal in length.

Another Way

Triangles can be described by the types of angles they have.

Draw a line to match the description of the triangle(s).

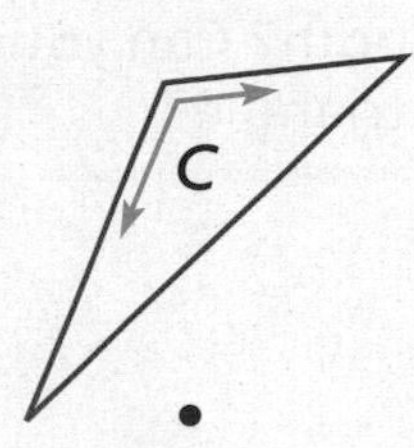

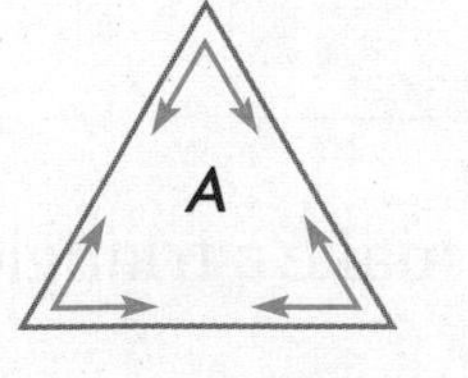

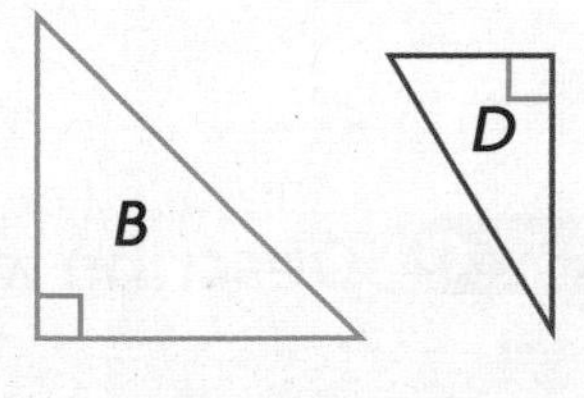

One angle is a right angle.

One angle is greater than a right angle.

Three angles are less than a right angle.

Math Talk MATHEMATICAL PRACTICES 2

Use Reasoning Can a triangle have two right angles?

Name ______________________________

Share and Show

1. Write the number of sides of equal length the triangle appears to have.

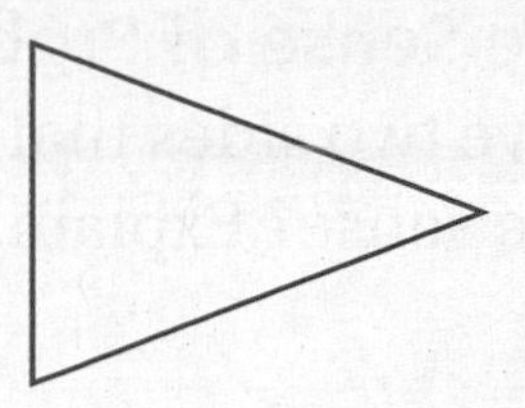

____

Use the triangles for 2–4. Write *F, G,* or *H*.

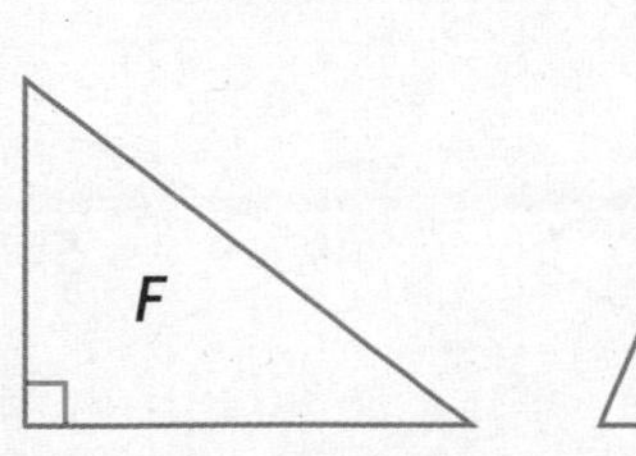

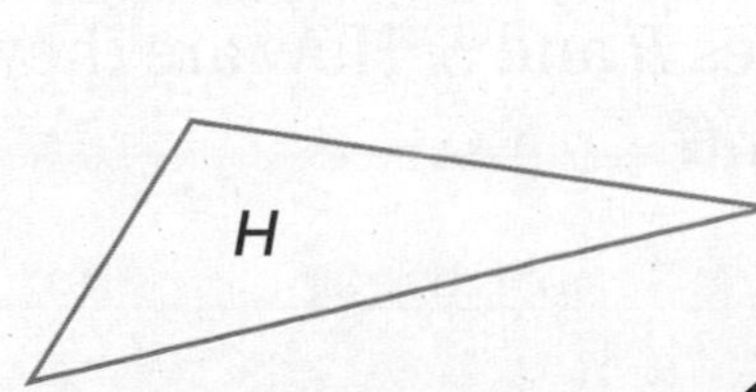

2. Triangle _____ has 1 right angle.

✓3. Triangle _____ has 1 angle greater than a right angle.

✓4. Triangle _____ has 3 angles less than a right angle.

Math Talk MATHEMATICAL PRACTICES 8

Generalize Explain the ways you can describe a triangle.

On Your Own

Use the triangles for 5–7. Write *K, L,* or *M*. Then complete the sentences.

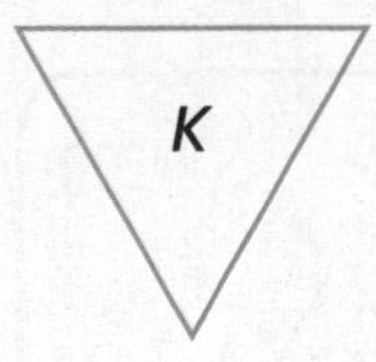

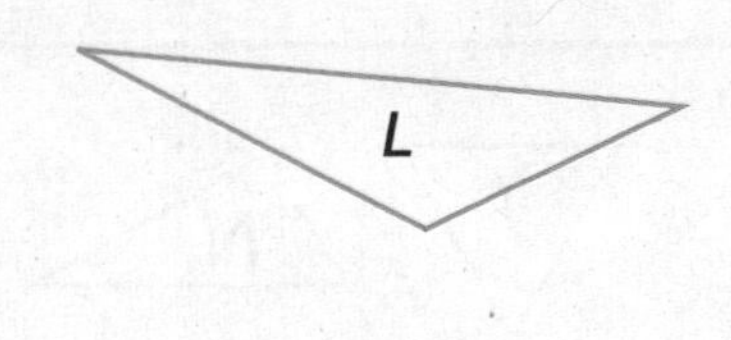

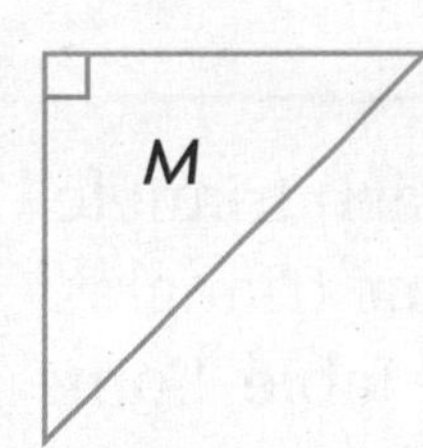

5. Triangle _____ has 1 right angle and appears to have _____ sides of equal length.

6. Triangle _____ has 3 angles less than a right angle and appears to have _____ sides of equal length.

7. Triangle _____ has 1 angle greater than a right angle and appears to have _____ sides of equal length.

Problem Solving • Applications 

8. MATHEMATICAL PRACTICE 1 **Make Sense of Problems** Martin said a triangle can have two sides that are parallel. Does his statement make sense? Explain.

____

____

9. GO DEEPER Compare Triangles *R* and *S*. How are they alike? How are they different?

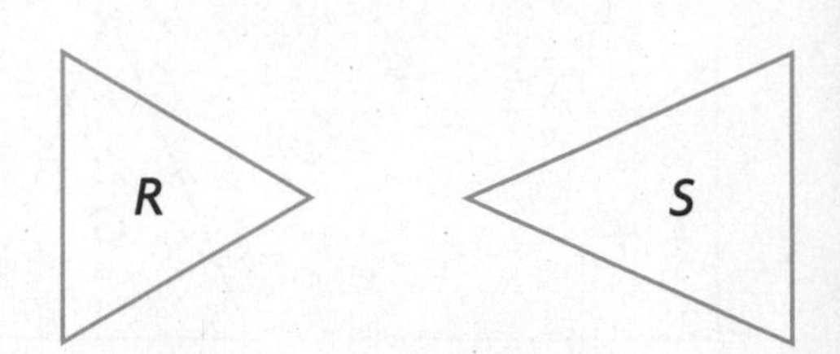

____

____

____

10. THINK SMARTER Use a ruler to draw a straight line from one corner of this rectangle to the opposite corner. What shapes did you make? What do you notice about the shapes?

____

____

11. THINK SMARTER Write the name of each triangle where it belongs in the table. Some triangles might belong in both parts of the table. Some triangles might not belong in either part.

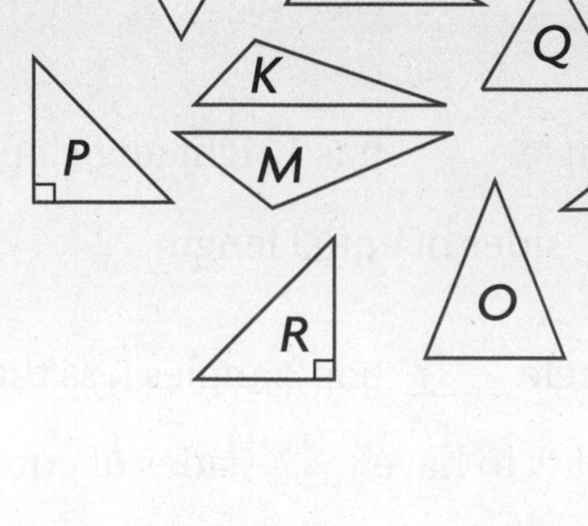

Has 1 Right Angle	Has at Least 2 Sides of Equal Length

Name ______________________________

Describe Triangles

Practice and Homework
Lesson 12.7

Common Core **COMMON CORE STANDARD—3.G.A.1**
Reason with shapes and their attributes.

Use the triangles for 1–3. Write *A*, *B*, or *C*. Then complete the sentences.

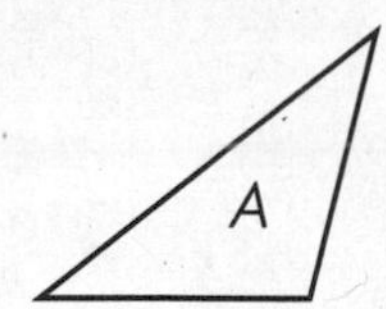

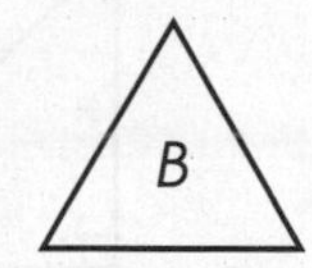

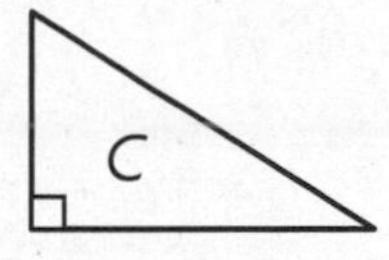

1. Triangle __B__ has 3 angles less than a right angle and appears to have __3__ sides of equal length.

2. Triangle ______ has 1 right angle and appears to have ______ sides of equal length.

3. Triangle ______ has 1 angle greater than a right angle and appears to have ______ sides of equal length.

Problem Solving

4. Matthew drew the back of his tent. How many sides appear to be of equal length?

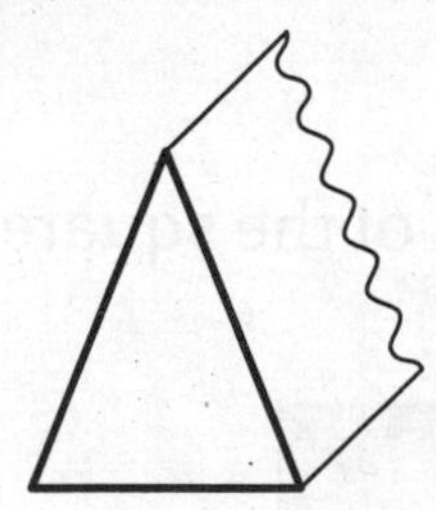

____

5. Sierra made the triangular picture frame shown. How many angles are greater than a right angle?

____

6. WRITE *Math* Draw a triangle that has two sides of equal length and one right angle.

____

____

Lesson Check (3.G.A.1)

1. How many angles less than a right angle does this triangle have?

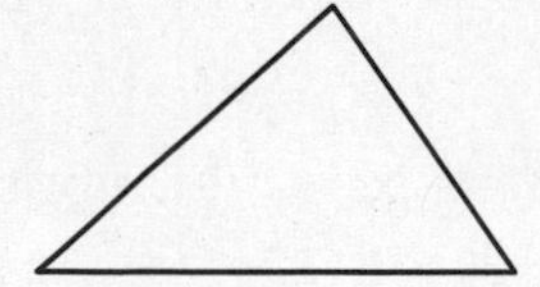

2. How many sides of equal length does this triangle appear to have?

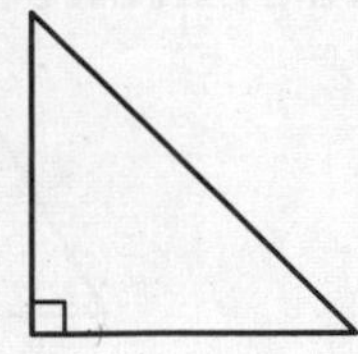

Spiral Review (3.NF.A.1, 3.MD.D.8, 3.G.A.1)

3. A quadrilateral has 4 right angles and 2 pairs of opposite sides that are parallel. What quadrilateral could it be?

4. Mason drew a quadrilateral with only one pair of opposite sides that are parallel. What quadrilateral did Mason draw?

5. What are the side lengths of a rectangle that has an area of 8 square units and a perimeter of 12 units?

6. What fraction of the square is shaded?

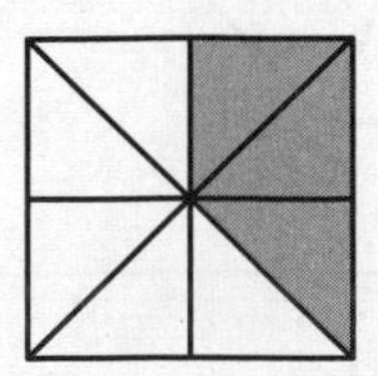

FOR MORE PRACTICE GO TO THE **Personal Math Trainer**

Name ______________________________

Problem Solving • Classify Plane Shapes

Essential Question How can you use the strategy *draw a diagram* to classify plane shapes?

Common Core Geometry—3.G.A.1

MATHEMATICAL PRACTICES
MP1, MP2, MP4, MP7

Unlock the Problem Real World

A **Venn diagram** shows how sets of things are related. In the Venn diagram at the right, one circle has shapes that are rectangles. Shapes that are rhombuses are in the other circle. The shapes in the section where the circles overlap are both rectangles and rhombuses.

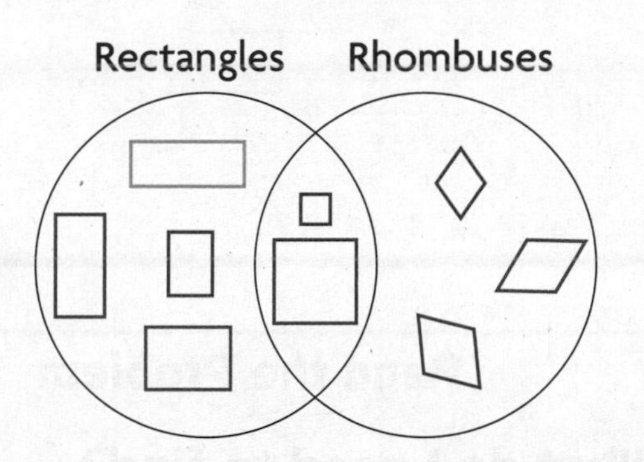

What type of quadrilateral is in both circles?

Read the Problem	Solve the Problem
What do I need to find? ______________________ ______________________	What is true about all quadrilaterals? ______________________ Which quadrilaterals have 2 pairs of opposite sides that are parallel? ______________________
What information do I need to use? the circles labeled __________ and __________	Which quadrilaterals have 4 sides of equal length? __________ Which quadrilaterals have 4 right angles? ______________________
How will I use the information? ______________________ ______________________ ______________________	The quadrilaterals in the section where the circles overlap have _____ pairs of opposite sides that are parallel, _____ sides of equal length, and _____ right angles. So, __________ are in both circles.

Math Talk MATHEMATICAL PRACTICES 1

Make Sense of Problems Does a △ fit in the Venn diagram? Explain.

Try Another Problem

The Venn diagram shows the shapes Andrea used to make a picture. Where would the shape shown below be placed in the Venn diagram?

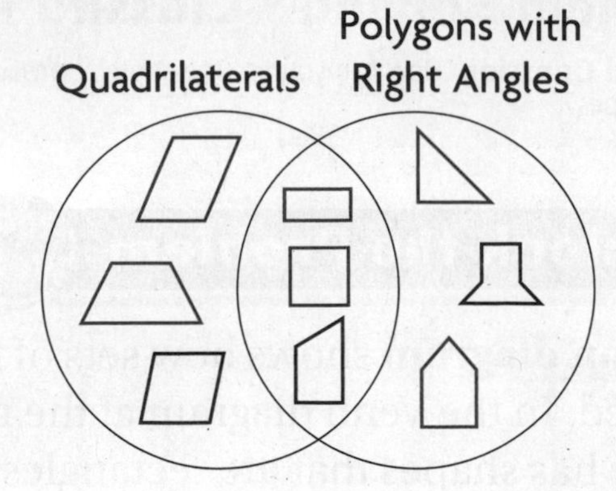

Read the Problem	Solve the Problem
What do I need to find?	**Record the steps you used to solve the problem.**
What information do I need to use?	
How will I use the information?	

1. How many shapes do not have right angles?

____

2. How many red shapes have right angles but are not quadrilaterals? ______________

3. MATHEMATICAL PRACTICE 2 **Reason Abstractly** What is a different way to sort the shapes?

____

____

Math Talk MATHEMATICAL PRACTICES 1

Make Sense of Problems What name can be used to describe all the shapes in the Venn diagram? Explain how you know.

Name ______________________________

Share and Show MATH BOARD

Use the Venn diagram for 1–3.

1. Jordan is sorting the shapes at the right in a Venn diagram. Where does the ◇ go?

   First, look at the sides and angles of the polygons.

   Next, draw the polygons in the Venn diagram.

   The shape has _____ sides of equal length

   and _____ right angles.

   So, the shape goes in the

   ____.

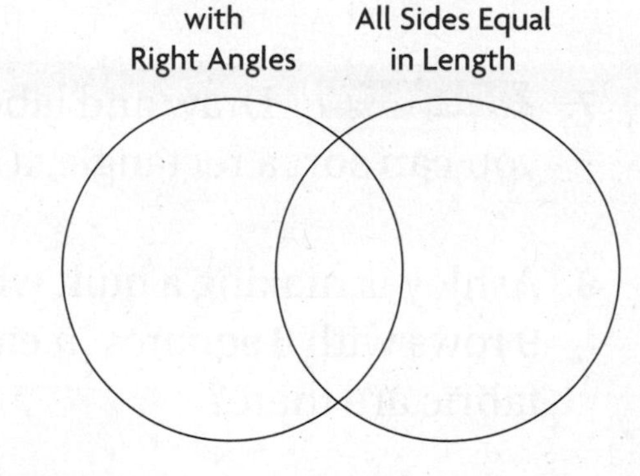

2. Where would you place a ⌂?

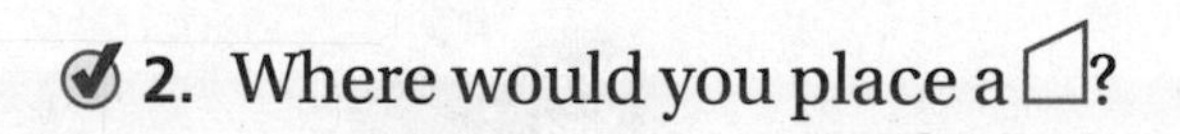

____

3. What if Jordan sorted the shapes by Polygons with Right Angles and Polygons with Angles Less Than a Right Angle? Would the circles still overlap? Explain.

____

____

4. GO DEEPER Eva drew the Venn diagram below. What labels could she have used for the diagram?

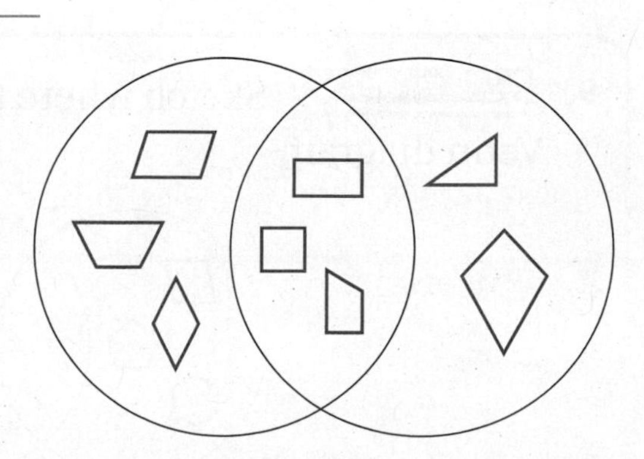

On Your Own

5. Ben and Marta are both reading the same book. Ben has read $\frac{1}{3}$ of the book. Marta has read $\frac{1}{4}$ of the book. Who has read more? ________

6. MATHEMATICAL PRACTICE 2 **Represent a Problem** There are 42 students from 6 different classes in the school spelling bee. Each class has the same number of students in the spelling bee. Use the bar model to find how many students are from each class.

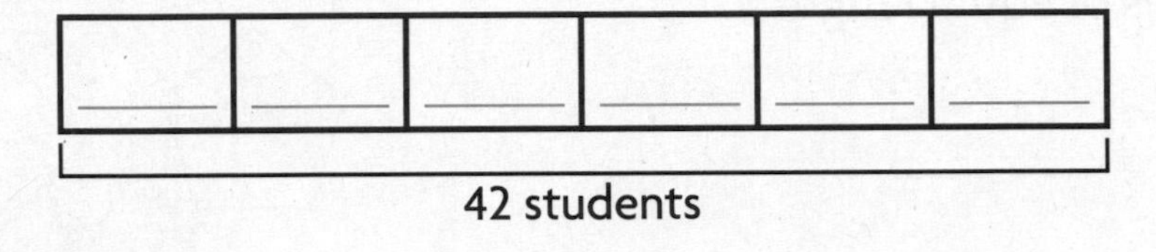

42 students

______ students ÷ ______ classes = ______ students

7. THINKSMARTER Draw and label a Venn diagram to show one way you can sort a rectangle, a square, a trapezoid, and a rhombus.

8. Ashley is making a quilt with squares of fabric. There are 9 rows with 8 squares in each row. How many squares of fabric are there?

____

Personal Math Trainer

9. THINKSMARTER+ Sketch where to place these shapes in the Venn diagram.

Polygons with All Sides of Equal Length

Quadrilaterals with Right Angles

Name ____________________

Problem Solving • Classify Plane Shapes

Common Core **COMMON CORE STANDARD—3.G.A.1**
Reason with shapes and their attributes.

Solve each problem.

1. Steve drew the shapes below. Write the letter of each shape where it belongs in the Venn diagram.

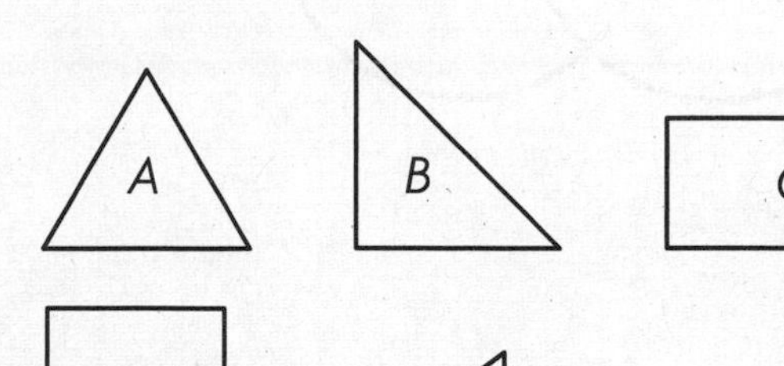

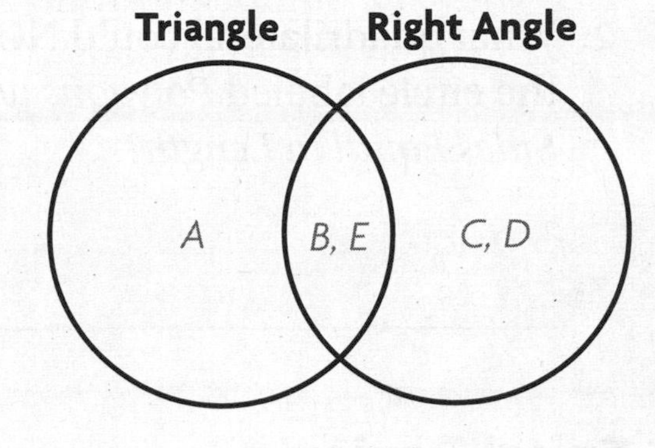

2. Janice drew the shapes below. Write the letter of each shape where it belongs in the Venn diagram.

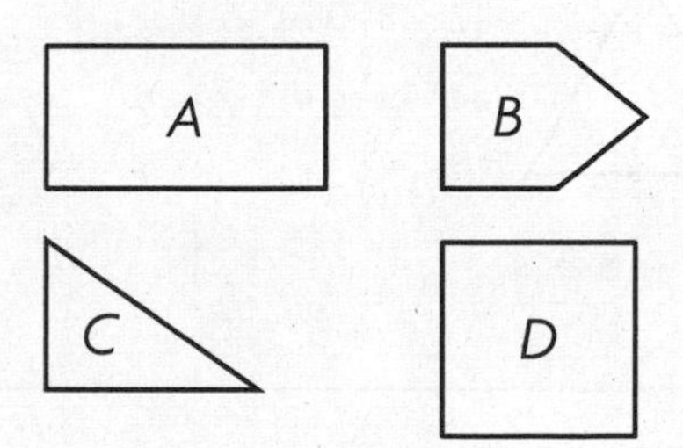

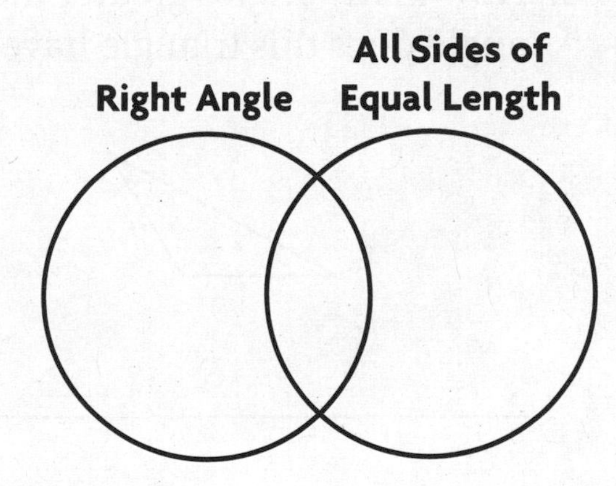

3. **WRITE** *Math* Draw a Venn diagram with one circle labeled *Quadrilaterals* and the other circle labeled *Polygons with More Than 3 Sides*. Draw at least two shapes in each section of the diagram. Explain why you drew the shapes you chose in the overlapping section

____

____

Lesson Check (3.G.A.1)

1. What shape would go in the section where the two circles overlap?

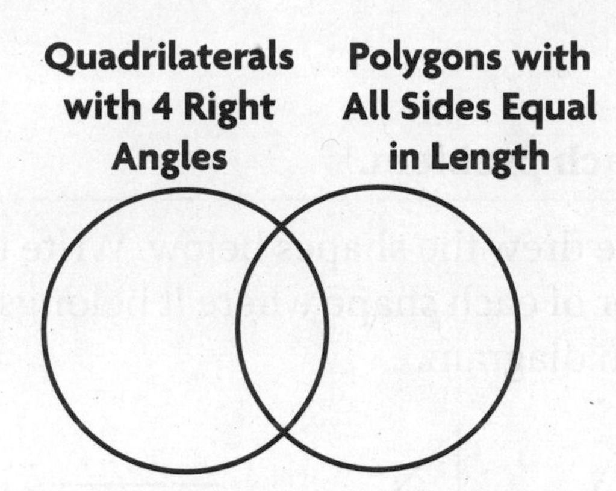

____

2. What quadrilateral could NOT go in the circle labeled *Polygons with All Sides Equal in Length*?

____

Spiral Review (3.NF.A.1, 3.G.A.1)

3. How many angles greater than a right angle does this triangle have?

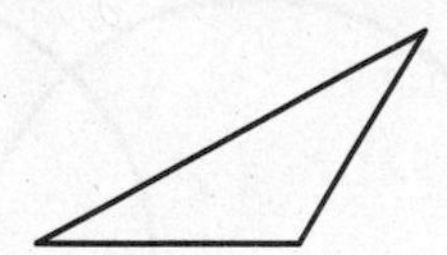

____

4. How many sides of equal length does this triangle appear to have?

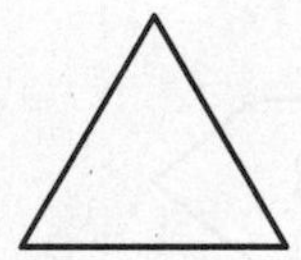

____

5. Madison drew this shape. How many angles less than a right angle does it have?

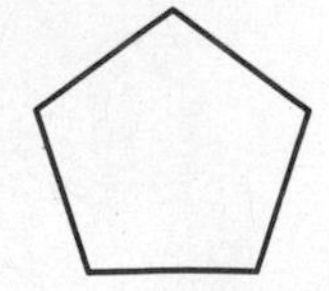

____

6. How many dots are in $\frac{1}{2}$ of this group?

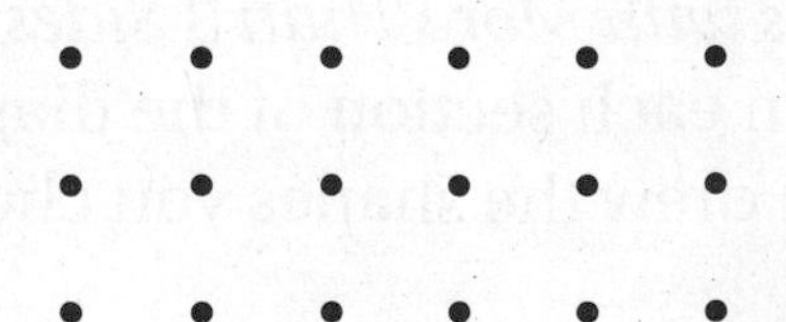

____

FOR MORE PRACTICE GO TO THE **Personal Math Trainer**

Name ______________________________

Relate Shapes, Fractions, and Area

Essential Question How can you divide shapes into parts with equal areas and write the area as a unit fraction of the whole?

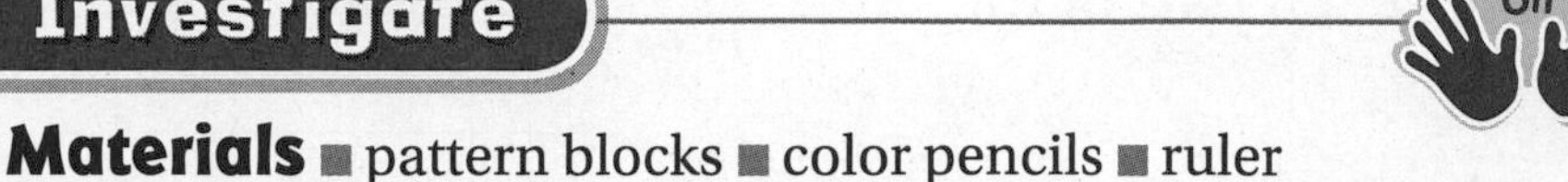

Common Core **Geometry—3.G.A.2**
Also 3.NF.A.1, 3.NF.A.3d, 3.MD.C.5

MATHEMATICAL PRACTICES
MP4, MP6, MP7, MP8

Investigate

Hands On

Materials ■ pattern blocks ■ color pencils ■ ruler

CONNECT You can use what you know about combining and separating plane shapes to explore the relationship between fractions and area.

A. Trace a hexagon pattern block.

B. Divide your hexagon into two parts with equal area.

C. Write the names of the new shapes. __________

D. Write the fraction that names each part of the whole you divided. ____
Each part is $\frac{1}{2}$ of the whole shape's area.

E. Write the fraction that names the whole area. ____

Math Idea
Equal parts of a whole have equal area.

Draw Conclusions

1. Explain how you know the two shapes have the same area.

____

____

2. Predict what would happen if you divide the hexagon into three shapes with equal area. What fraction names the area of each part of the divided hexagon? What fraction names the whole area?

____

3. THINKSMARTER Show how you can divide the hexagon into four shapes with equal area.

Each part is ____ of the whole shape's area.

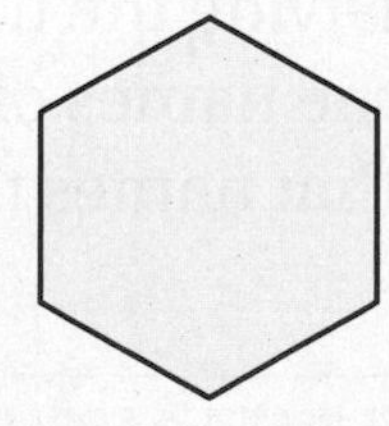

Make Connections

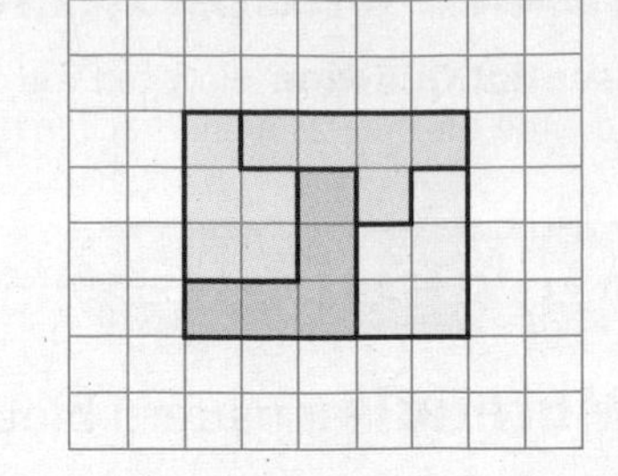

The rectangle at the right is divided into four parts with equal area.

- Write the unit fraction that names each part of the divided whole. _____
- What is the area of each part? _____
- How many $\frac{1}{4}$ parts does it take to make one whole? _____
- Is the shape of each of the $\frac{1}{4}$ parts the same? _____
- Is the area of each of the $\frac{1}{4}$ parts the same? Explain how you know.

____

____

Divide the shape into equal parts.

Draw lines to divide the rectangle below into six parts with equal area.

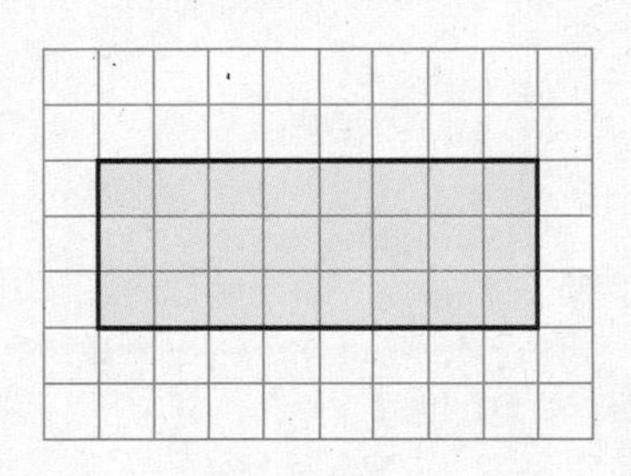

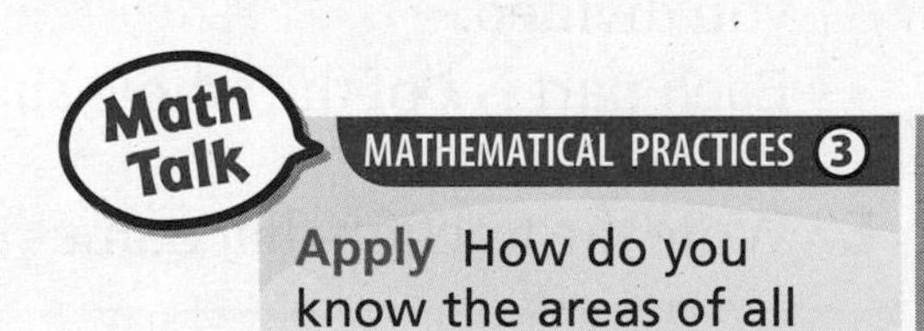

- Write the fraction that names each part of the divided whole. _____
- Write the area of each part. _________
- Each part is _____ of the whole shape's area.

Share and Show

1. Divide the trapezoid into 3 parts with equal area. Write the names of the new shapes. Then write the fraction that names the area of each part of the whole.

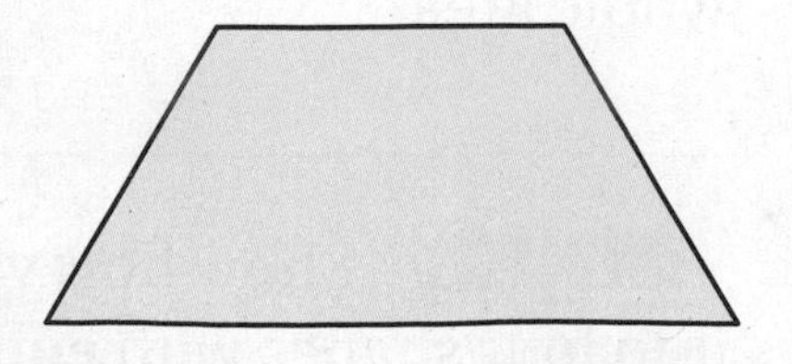

____

Name ______________________________

Draw lines to divide the shape into equal parts that show the fraction given.

2. 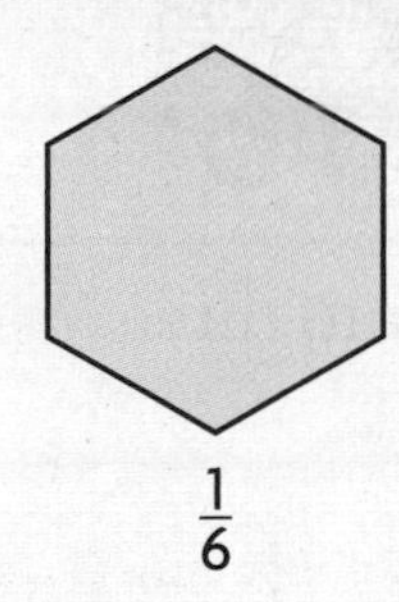

$\frac{1}{6}$

3. 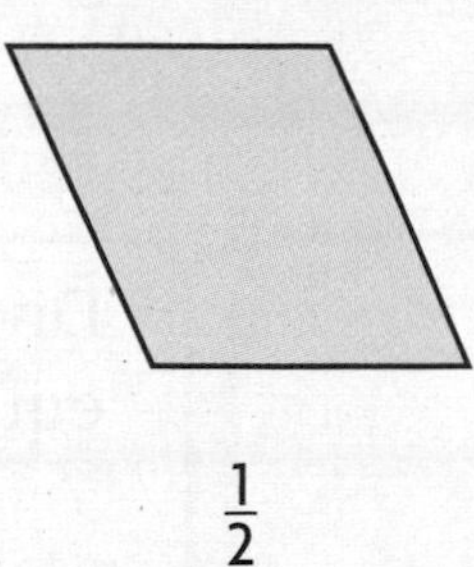

$\frac{1}{2}$

✓4. 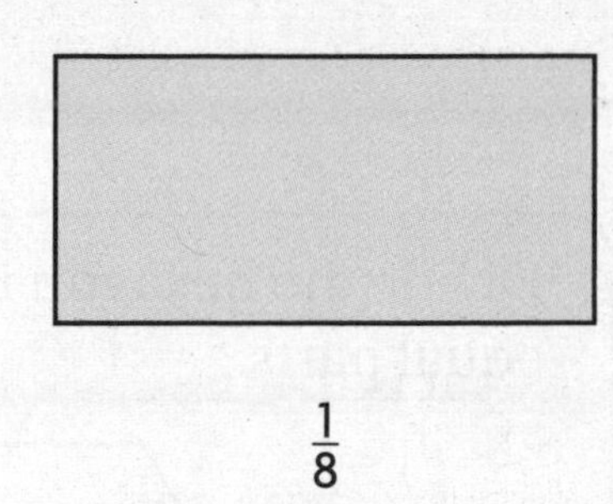

$\frac{1}{8}$

Draw lines to divide the shape into parts with equal area. Write the area of each part as a unit fraction.

5. 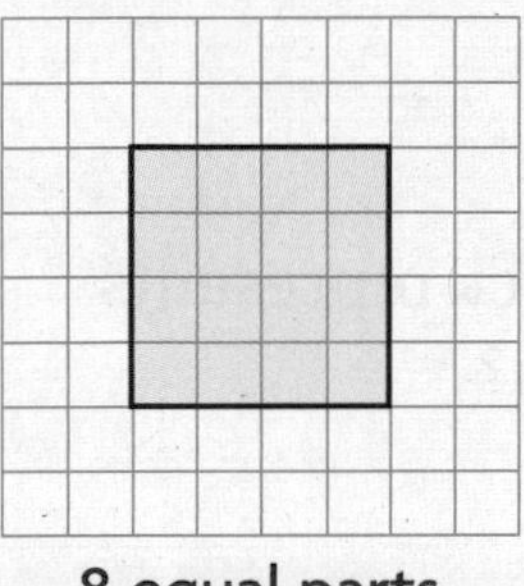

8 equal parts

____

✓6. 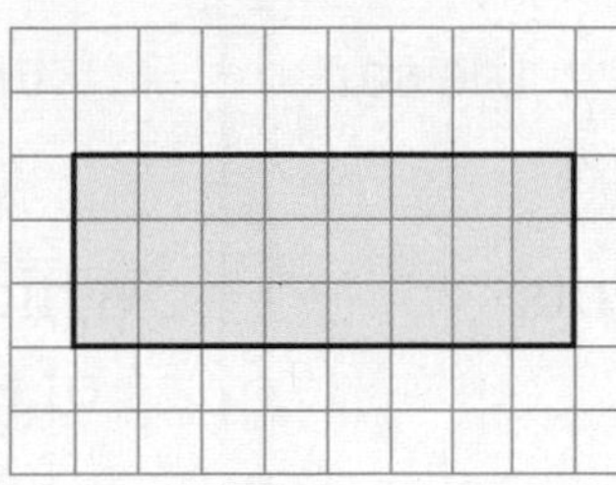

6 equal parts

____

7. 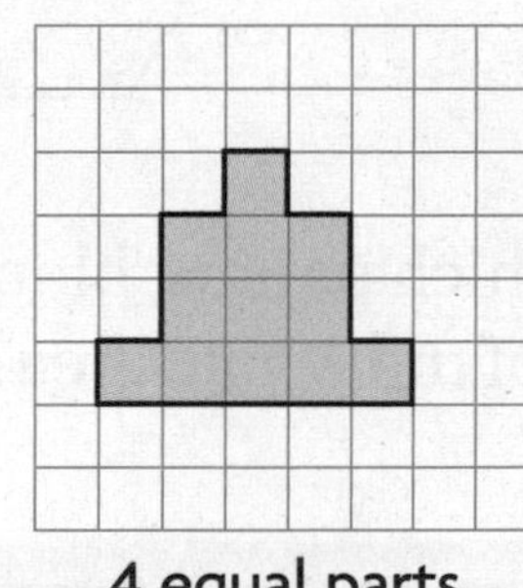

4 equal parts

____

Problem Solving • Applications

8. MATHEMATICAL PRACTICE ② **Use Reasoning** If the area of three is equal to the area of one , the area of how many equals four ? Explain your answer.

____

9. THINK SMARTER Divide each shape into the number of equal parts shown. Then write the fraction that describes each part of the whole.

2 equal parts

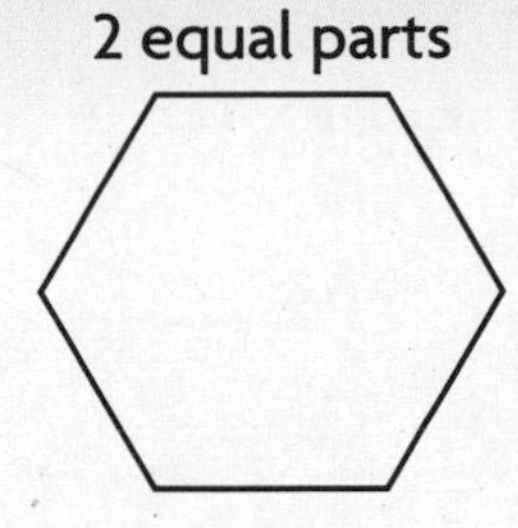

____

4 equal parts

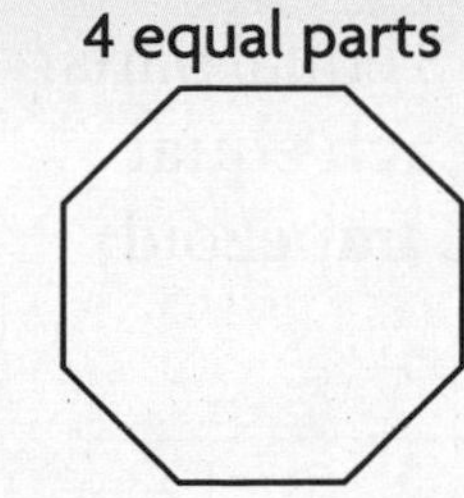

____

6 equal parts

____

10. THINK SMARTER **Sense or Nonsense?**

Divide the hexagon into six equal parts.

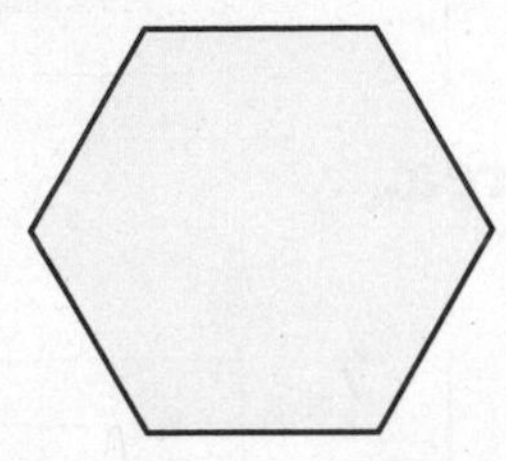

Which pattern block represents $\frac{1}{6}$ of the whole area?

____

Divide the trapezoid into three equal parts.

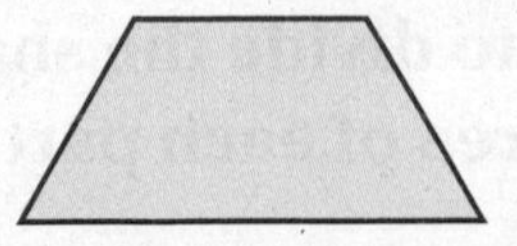

Which pattern block represents $\frac{1}{3}$ of the whole area?

____

Alexis said the area of $\frac{1}{3}$ of the trapezoid is greater than the area of $\frac{1}{6}$ of the hexagon because $\frac{1}{3} > \frac{1}{6}$. Does her statement make sense? Explain your answer.

____

____

____

____

- Write a statement that makes sense.

  ____

  ____

- GO DEEPER What if you divide the hexagon into 3 equal parts? Write a sentence that compares the area of each equal part of the hexagon to each equal part of the trapezoid.

  ____

  ____

Name ______________________________

Relate Shapes, Fractions, and Area

Common Core **COMMON CORE STANDARD—3.G.A.2**
Reason with shapes and their attributes.

Draw lines to divide the shape into equal parts that show the fraction given.

1.

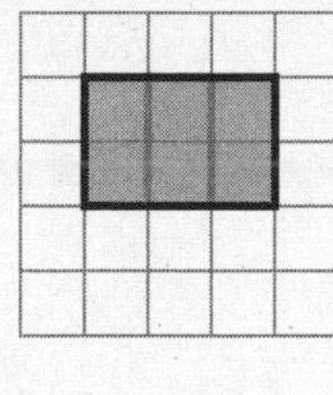

$\frac{1}{3}$

2.

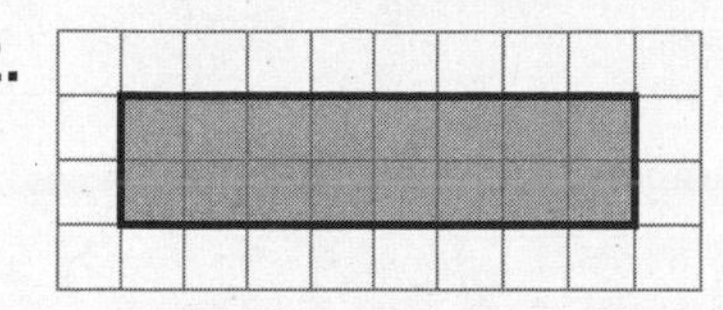

$\frac{1}{8}$

3.

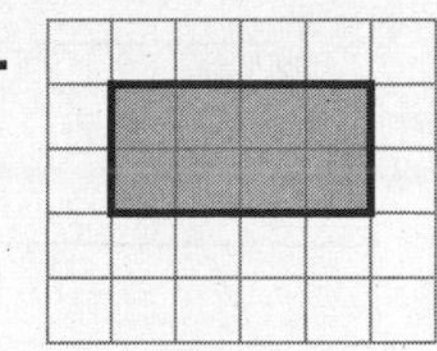

$\frac{1}{2}$

Draw lines to divide the shape into parts with equal area. Write the area of each part as a unit fraction.

4.

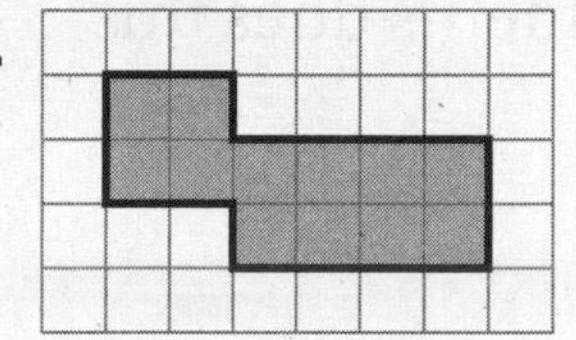

4 equal parts

______

5.

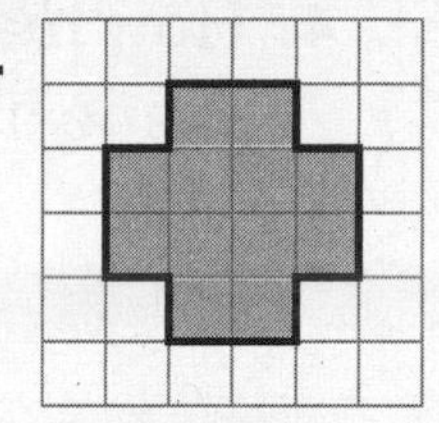

6 equal parts

______

6.

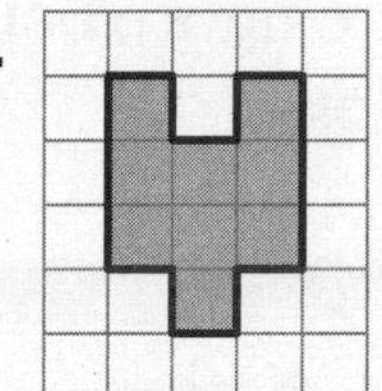

3 equal parts

______

Problem Solving

7. Robert divided a hexagon into 3 equal parts. Show how he might have divided the hexagon. Write the fraction that names each part of the whole you divided.

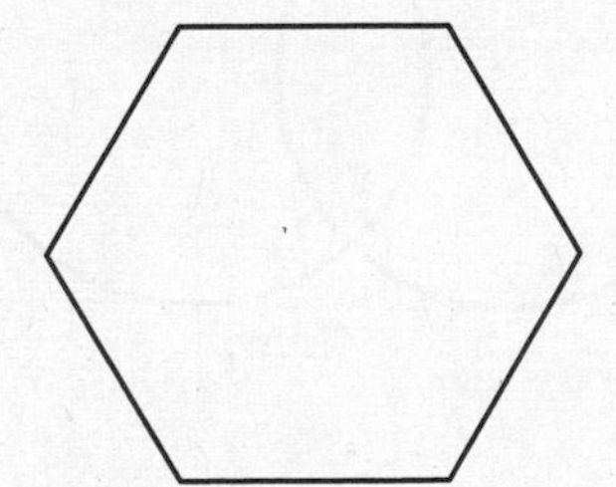

____

8. WRITE *Math* Trace a pattern block. Divide it into two equal parts, and write a unit fraction to describe the area of each part. Explain your work.

____

Lesson Check (3.G.A.1)

1. What fraction names each part of the divided whole?

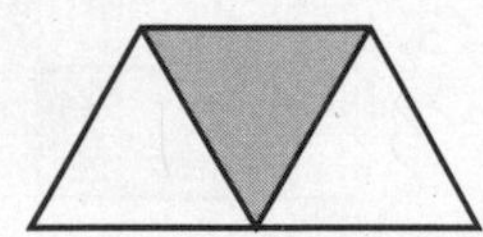

2. What fraction names the whole area that was divided?

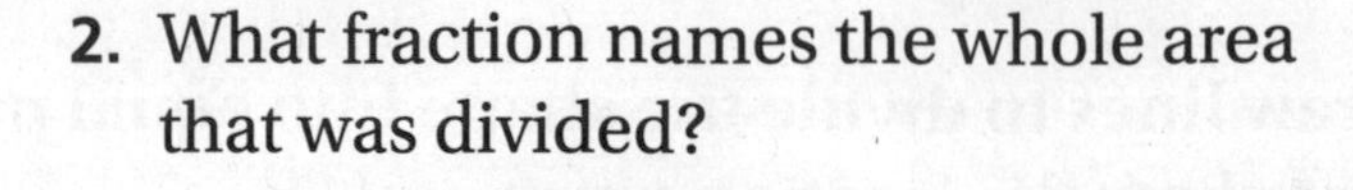

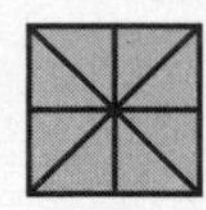

Spiral Review (3.G.A.1)

3. Lil drew the figure below. Is the shape open or closed?

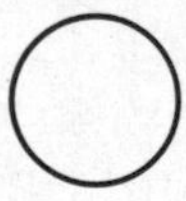

4. How many line segments does this shape have?

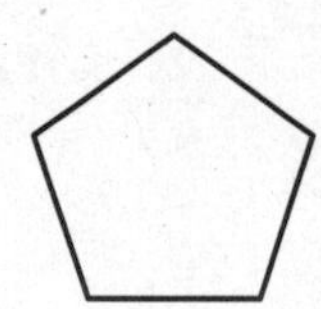

Use the Venn diagram for 5–6.

5. Where would a square be placed in the Venn diagram?

6. Where would a rectangle be placed in the Venn diagram?

Right Angle **All Sides of Equal Length**

FOR MORE PRACTICE GO TO THE **Personal Math Trainer**

Name ______________________________

Chapter 12 Review/Test

1. Which words describe this shape? Mark all that apply.

Ⓐ polygon

Ⓑ open shape

Ⓒ pentagon

Ⓓ quadrilateral

2. Umberto drew one side of a quadrilateral with 4 equal sides and no right angles. Draw the other 3 sides to complete Umberto's shape.

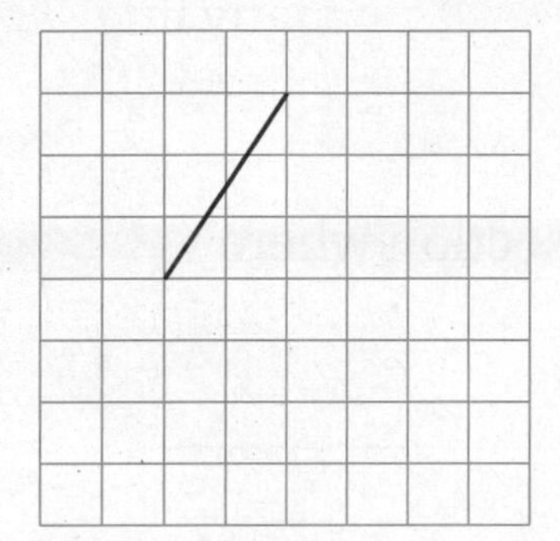

3. Mikael saw a painting that included this shape.

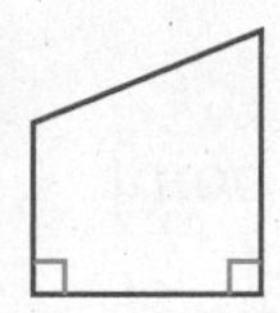

For numbers 3a–3d, select True or False for each statement about the shape.

3a.	The shape has no right angles.	○ True	○ False
3b.	The shape has 2 angles greater than a right angle.	○ True	○ False
3c.	The shape has 2 right angles.	○ True	○ False
3d.	The shape has 1 angle greater than a right angle.	○ True	○ False

GO DIGITAL Assessment Options Chapter Test

4. GO DEEPER Fran used a Venn Diagram to sort shapes.

Part A

Draw another plane shape that belongs inside the left circle of the diagram but NOT in the section where the circles overlap.

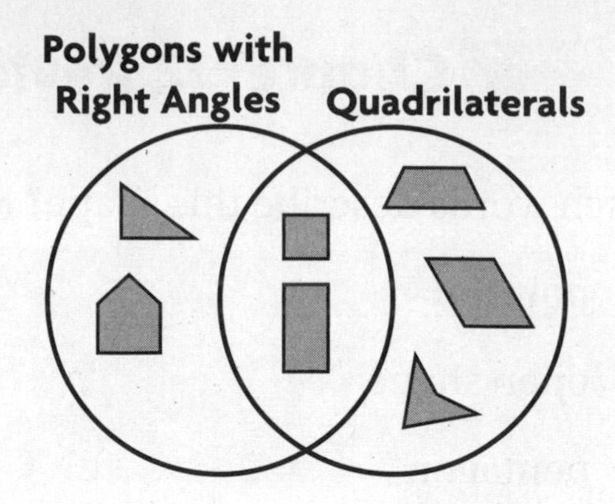

Part B

How can you describe the shapes in the section where the circles overlap?

____

____

5. Match each object in the left column with its name in the right column.

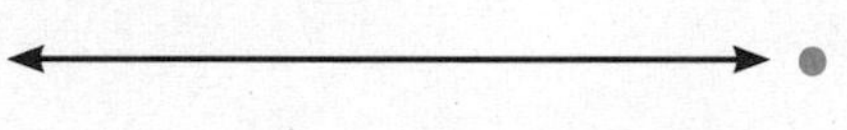 • point

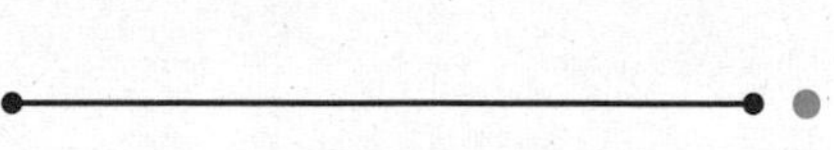 • line

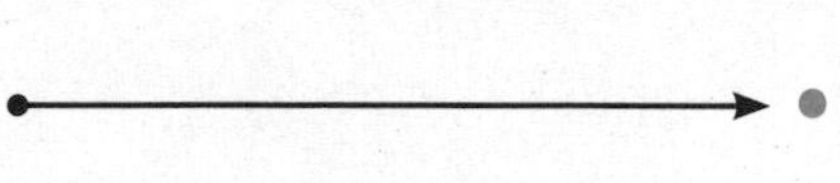 • ray

 • line segment

6. Describe the angles and sides of this triangle.

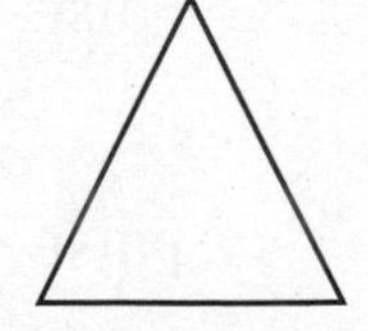

____

Name ___________________________________

7. Which words describe this shape. Mark all that apply.

rectangle Ⓐ rhombus Ⓑ quadrilateral Ⓒ square Ⓓ

8. Divide each shape into the number of equal parts shown. Then write the fraction that describes each part of the whole.

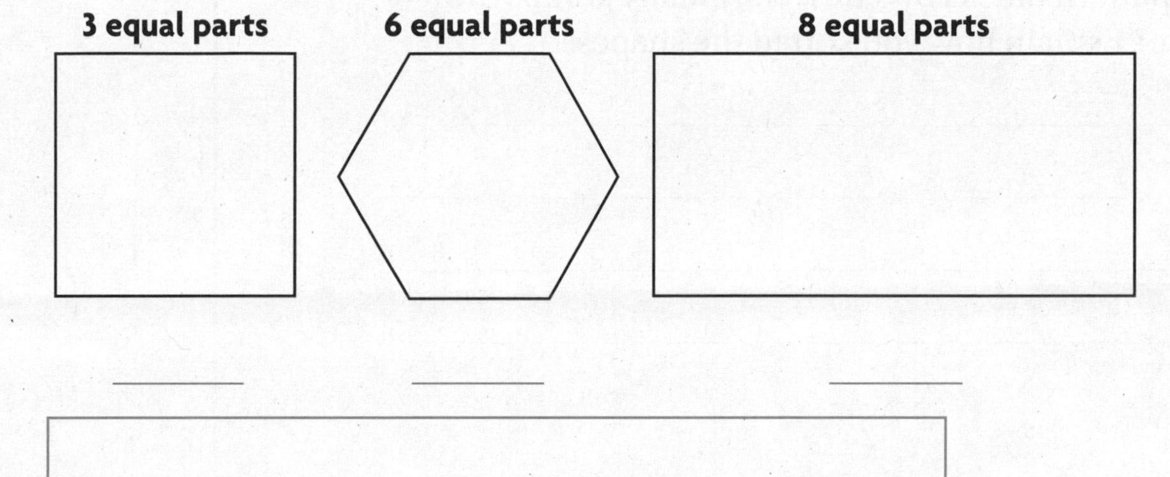

_______ _______ _______

9. Han drew a triangle with 1 angle greater than a right angle.

For numbers 9a–9d, choose Yes or No to tell whether the triangle could be the triangle Han drew.

9a. 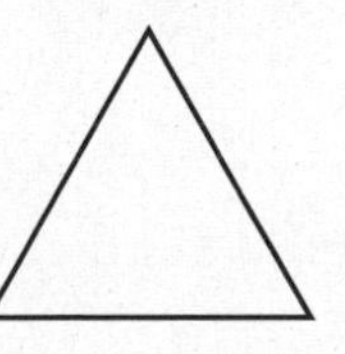 ○ Yes ○ No

9b. 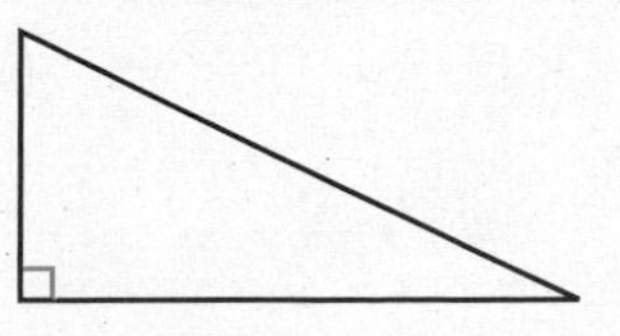 ○ Yes ○ No

9c. 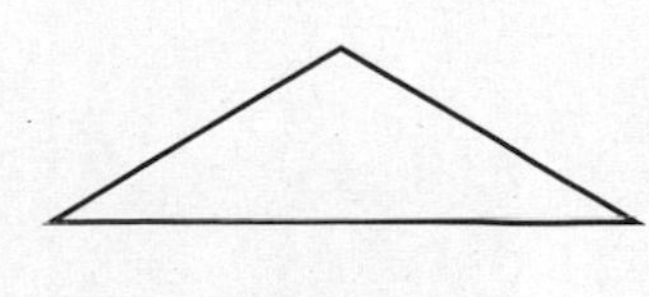 ○ Yes ○ No

9d. ○ Yes ○ No

Personal Math Trainer

10. THINK SMARTER + Look at this group of pattern blocks.

Part A

Sort the pattern blocks by sides. How many groups did you make? Explain how you sorted the shapes.

____

____

____

____

Part B

Sort the pattern blocks by angles. How many groups did you make? Explain how you sorted the shapes.

____

____

____

____

11. Teresa drew a quadrilateral that had 4 sides of equal length and no right angles. What quadrilateral did she draw?

____

Name ______________________________

12. Rhea used a Venn diagram to sort shapes. What label could she use for circle A?

A **Polygons with All Sides of Equal Length**

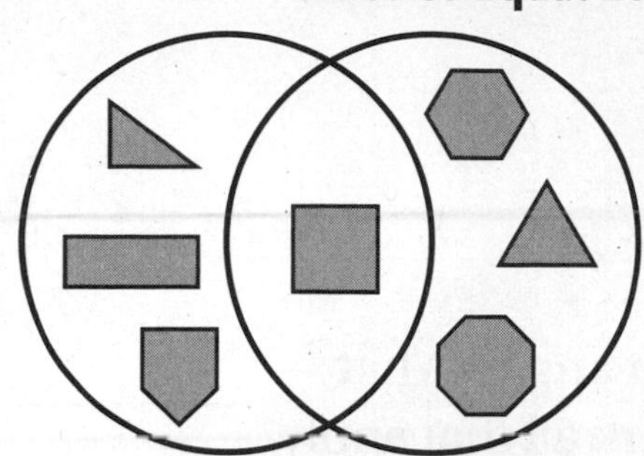

____

13. Colette drew lines to divide a rectangle into equal parts that each represent $\frac{1}{6}$ of the whole area. Her first line is shown. Draw lines to complete Colette's model.

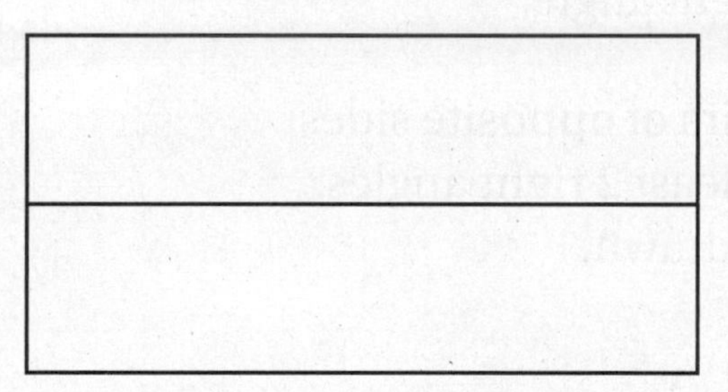

14. Brad drew a quadrilateral. Select the pairs of sides that appear to be parallel. Mark all that apply.

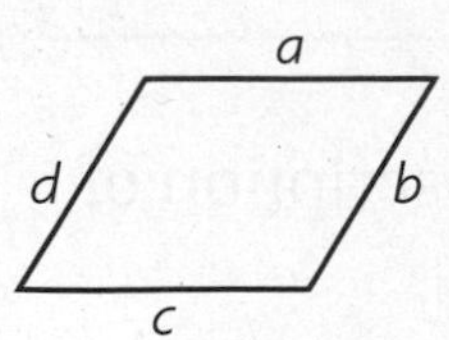

Ⓐ a and b

Ⓑ b and d

Ⓒ c and a

Ⓓ d and c

15. Give two reasons that this shape is **not** a polygon.

____

16. A triangle has 1 angle greater than a right angle. What must be true about the other angles? Mark all that apply.

- Ⓐ At least one must be less than a right angle.
- Ⓑ One could be a right angle.
- Ⓒ Both must be less than a right angle.
- Ⓓ One must be greater than a right angle.

17. Ava drew a quadrilateral with 2 pairs of opposite sides that are parallel. The shape has at least 2 right angles. Draw a shape that Ava could have drawn.

18. For 18a–18d, select True or False for each description of a ray.

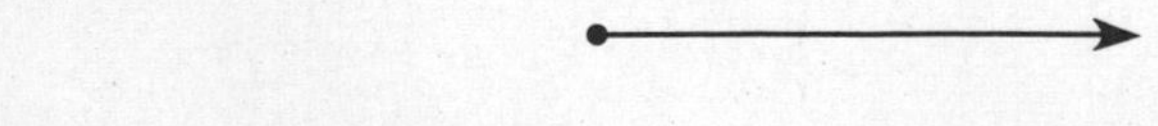

18a.	straight	○ True	○ False
18b.	has 2 endpoints	○ True	○ False
18c.	part of a line	○ True	○ False
18d.	continues in 1 direction	○ True	○ False